Weather Forecaster to Research Scientist

Weather Forecaster to Research Scientist

My Career in Meteorology

A Memoir

Robert M. Atlas

AMERICAN METEROLOGICAL SOCIETY

Cover images:
Photo of Robert M. Atlas courtesy of NOAA AOML.
Map figure from R. Atlas, M. Ghil, and M. Halem, "The Effect of Model Resolution and Satellite Sounding Data on GLAS Model Forecasts", Monthly Weather Review 110, 7 (1982): 662-682, https://doi.org/10.1175/1520-0493(1982)110<0662:TEOMRA>2.0.CO;2

Published by the American Meteorological Society
45 Beacon Street, Boston, Massachusetts 02108

The mission of the American Meteorological Society is to advance the atmospheric and related sciences, technologies, applications, and services for the benefit of society. Founded in 1919, the AMS has a membership of more than 13,000 and represents the premier scientific and professional society serving the atmospheric and related sciences. Additional information regarding society activities and membership can be found at www.ametsoc.org.

Library of Congress Control Number: 2021948671

Print ISBN: 978-1-944970-77-2
eISBN: 978-1-944970-78-9

Contents

Foreword

Dave Jones

When I was six years old, I became fascinated with the weather. I saw lightning strike a tree across the street from my house and witnessed ball lightning float across the road toward my window before simply vanishing. I was hooked. From then on, I read as much as possible about the weather, started keeping logs of the high and low temperatures, and asked my parents for a weather station and a police scanner for my birthday so I could listen to see if there was any damage after passing thunderstorms. At 12 years old and in seventh grade, I was offered the opportunity to take an earth science class in meteorology at the local community college. But it wasn't enough.

In high school, I took an evening weather forecasting class back at that same community college in Catonsville, Maryland. It wasn't for credit—just a class about weather forecasting. I would stay late and ask the teacher questions about the maps he brought to class; then I would ask him if I could take them home. Hanging the maps in an old photography "darkroom" in my parents' house, I created my "weather center."

The teacher of that weather forecasting class was Bob Atlas. At the time, he was a research scientist who worked at NASA's Goddard Space Flight Center (GSFC) in Greenbelt, Maryland, and he would drive at least 40 minutes to teach that class one night per week. I found out later that he loved weather probably more than I did. He must have seen something in my enthusiasm, my desire to have every map, and my determination to be the first one in the classroom for each class. He became my mentor, my advisor, and my friend. It's been just about 40 years since that weather forecasting class, and I have much to thank Bob Atlas for.

On the last day of class Bob asked me where I was going to college. I had applied to Penn State, had been accepted, but couldn't afford it. So I told him I was going to Catonsville Community College for my first two years because my mom worked there and I could get reimbursed for tuition if I received a grade of C or better. Without hesitation, Bob said, "NASA Goddard Space Flight Center hires interns, but you would have to go to the University of Maryland in College Park." After two years at the community college, I applied to the University of Maryland, got accepted, and NASA offered me a paid internship to work in the Global Modeling and Simulation Branch in Building 22 at NASA Goddard.

For two years, Bob took me under his wing and taught me about the dynamics of weather systems and especially winter storms. I eventually got to perform satellite retrievals using the McIDAS workstation. I worked late at NASA and even went there to do my homework because I felt "smarter" there. Being surrounded by the brightest scientists on the planet made me a better meteorologist. I could walk up two flights of stairs and attend the map briefings with Paul Kocin, Louis Uccellini, Dr. Joanne Simpson, and Dr. David Atlas (one of the pioneers of radar meteorology).

Later, more opportunities due to Bob came. How many undergraduate students can say that they were offered internships at the National Hurricane Center and NBC in Washington, DC, during the same summer . . . while working at NASA? I felt truly blessed.

Bob was also one of my professors at the University of Maryland. He had a wonderful way of relating Pettersen's development equation and other equations to the weather maps. I certainly was not the best at math, but Bob's way of teaching made me want to understand the equations, and for that I am forever grateful.

When Bob described his early career experiences of working for the *AM America* and *Good Morning America* shows at ABC in New York, I was awed, mainly because of the huge responsibility of creating the forecasts for everyone around the nation to see on TV. But then I was asked to visit NBC WRC-TV in Washington, DC, by a classmate (who is now a longtime chief broadcast meteorologist in Jacksonville, Florida). I ended up interning with Bob Ryan (the first broadcast meteorologist to become president of the American Meteorological Society) in a position where I could take some of the cool research from NASA over to NBC, and every once in a while, Bob Ryan would put it on-air. It was awesome!

Thanks to Bob Atlas and the experience at NASA, I had great confidence that I could operate the weather graphics computers at NBC in Washington, DC. Soon I was helping to produce weather graphics that were used on-air every day while also working at NASA GSFC. Bob Atlas and I would engage in conversations about the different environments research meteorology and broadcast meteorology generated. I learned so much in the research environment but really liked the fast-paced nature of broadcast meteorology.

Again, the opportunities came, stemming from Bob Atlas's mentoring and the opportunities he gave me. I was hired by a weather computer graphics company and ended up training broadcast meteorologists all around the nation. And then one day, after being invited by Bob Ryan to make an audition tape for fun, I was offered a job as an on-air meteorologist to launch NBC4's new weekend morning show in Washington, DC. For almost 10 years I got to live the dream of talking about the weather to millions of people and even joined the *Today Show* in New York on multiple occasions to deliver the weather and talk with people on the street in New York City. Thank you, Bob Atlas.

Today, I am CEO of StormCenter Communications, Inc., a company I founded in 2001. At least once per week, I think about that kid who just loved the weather (I still do). And I think about how grateful I am that the stars aligned on that fateful evening when I walked into that weather forecasting class at Catonsville Community College and met Dr. Bob Atlas. He is an amazing scientist and mentor, someone who has my utmost respect and someone whom I will always consider my friend.

In this memoir, Bob describes his career in meteorology, which spanned six decades. Dr. Atlas began his career at a time when satellite meteorology and operational numerical weather prediction were in their infancy. He gives examples of his experiences forecasting the weather and how this evolved. He describes his varied research for NASA, leading to (1) the first operationally accepted beneficial impact of quantitative satellite data on numerical weather prediction, (2) his pioneering work with satellite ocean surface winds, (3) his modifications to the methodology for observing system simulation experiments (OSSEs) that increased their realism and made OSSEs a much more useful tool for determining the potential impacts of modifications or additions to the global observing system, and (4) his research to improve understanding and prediction of extreme weather. Bob also describes his leadership of scientific organizations in NASA and NOAA, and his experiences teaching at several universities. I believe this memoir will prove useful to a very broad audience.

Preface

On March 2, 2019, I retired from the federal government after 58 years as a meteorologist. Forty-three of these years were as a federal employee, first in the Air Weather Service of the United States Air Force, then with NASA, and finally with NOAA. In the course of my career, I worked as an operational forecaster, an educator, a research scientist, a television meteorologist, and a leader of scientific research organizations. It is my hope that this memoir will be useful and inspiring to those who read it. To all of my friends and colleagues that I have worked with over the years, I want to say what a pleasure and privilege it has been to know you and work with you. To all students and young scientists in this field of endeavor, it is my distinct hope that you will have rewarding careers in which you contribute to both advancing scientific knowledge and saving lives.

This memoir is intended for a broad audience ranging from those in the general public with an interest in weather to students at a variety of levels, practicing weather forecasters, educators, research scientists, and science leaders. It represents my recollection of events throughout my career in meteorology. In some instances, my recollections may differ from others', but I have fact checked wherever possible and also include relevant references, and Appendix B lists all of my publications. Figures, maps, statistics, and other technical results related to the research I describe are contained within those references. In most places where meteorological terminology is introduced, I've attempted to include a brief nontechnical description of its meaning.

The memoir begins with my early years in meteorology, specifically when I became interested in the weather, and describes how this interest grew and evolved. It covers my education, my time as a weather officer in the U.S. Air Force, some of my research, and the leadership and teaching positions I have held.

My career spans a period of rapid development in meteorology. This includes forecasting at a time when it was still largely subjective, the early days of operational numerical weather prediction and its development to the present time, the launch of the first weather satellites, and the rapid growth of satellite meteorology.

The names of several of the organizations that I mention in this memoir changed over the period of time covered. In particular, the U.S. Weather Bureau became the National Weather Service; the National Meteorological Center became the National Centers for Environmental Prediction; the Goddard Laboratory for Atmospheric Sciences became the Goddard Laboratory for Atmospheres; and the Air Weather Service of the U.S. Air Force became the Air Force Weather Agency. In the memoir, I refer to these organizations by the name appropriate for the time being discussed.

I must point out that none of the successes that I've had in my career would have been possible if I had been working by myself. As will be seen throughout this memoir, there are many great meteorologists, oceanographers, mathematicians, computer programmers, and other scientists who have played a strong role throughout my career. Indeed, one of the characteristics of our field is that it draws so many outstanding scientists to unlock the mysteries of nature and improve our ability to predict it.

I cannot name all of the people who have had a positive influence on my career and my life here, but there are several that I would especially like to acknowledge. First and foremost, my wife Shirley, my son Ken, my daughter Robin, all of my grandchildren, and of course my parents. Next my teachers, especially Professors Benjamin Abell and Jerome Spar. Next, my students, who also became colleagues and friends, especially Bob Rosenberg, Lisa Baastians, Dave Jones, Joe Ardizzone, Joe Terry, Andy Pursch, Austin Conaty, Dennis Bungato, Steve Palm, Oreste Reale, Tom Cuff, and Ben Kirtman, to name a few. I'd also like to thank Gladys Medina, who served as my executive assistant at NOAA's Atlantic Oceanographic and Meteorological Laboratory (AOML). Interacting with her on a daily basis made my time there especially enjoyable. Finally, I'd like to acknowledge two of my earliest supervisors as a federal scientist, Drs. Milt Halem and Eugenia Kalnay. They each inspired me in my research in different ways, and from them I learned a great deal about leadership.

Robert M. Atlas
December 2021

Chapter 1

Early Years

I don't know exactly when I first became interested in meteorology, but I do remember my parents telling me that even as a very small child I would often be looking up at the sky, observing the clouds, and asking questions about the weather. In 1960, at age 12, I decided that it would be nice to keep a scrapbook of the weather. I wanted to be able to look back at precisely what the weather was on any particular date. As I cut out the weather pages from newspapers and perused the maps and observations, my interest grew and I knew that I needed to learn more about this fascinating subject. I searched our local bookstore and found a copy of the book *How to Know and Predict the Weather* (original title: *How About the Weather*) by Robert Moore Fisher.[1] This extraordinary book contained fascinating stories and a wealth of information about the atmosphere, weather, and weather forecasting. From this book, I learned about the layers of the atmosphere, the different types of clouds, and many interesting types of weather and weather phenomena. This included fronts, air masses, fog, hurricanes, and severe local storms.

About this time, I saw an exciting ad on television for a Lionel Weather Station. I remember my father being very angry with me for spending $19 on it, which was nearly all the money I had saved up to that time. I told him that this would be my career, but he didn't believe this until several years later. With the Lionel Weather Station, I was able to observe and record the weather, and I soon began forecasting the weather based on local observations of clouds,

1. Fisher, 1953.

pressure changes, and winds. This evolved into the "synoptic" method of forecasting as I began to make increasing use of surface weather maps and then surface and 500 mb (about 18,000 feet above sea level) constant pressure upper-air maps available from the U.S. Weather Bureau's Daily Weather Map series. These maps arrived in the mail every day but Sunday, and from them I would study the weather characteristics of the different air masses and fronts across the country.

At the start of the seventh grade in junior high school, I had not been a very good student. This was especially true in science and math. But once I became interested in weather and learned that I would need to graduate from college to become a meteorologist, everything changed. By the end of the seventh grade, I had become one of the best science and math students in the school and I was determined to learn all that I could to achieve my career goal.

The following year, one of my junior high school teachers introduced me to Mike Fayne, a substitute teacher at our school, whose main job was working as a forecaster for the U.S. Weather Bureau in New York City.[2] Mr. Fayne invited me to come for a tour of the forecast office, and while there I was asked if I would like to be an apprentice forecaster. For me, this was a dream come true, and I readily accepted. I spent five years there and learned a tremendous amount of both meteorology and practical forecasting from Mr. Fayne, as well as from three other very talented forecasters, Abe Zwaka, Joe Harrison, and Tom Grant.

Among other things, I learned how to use all of the constant pressure upper-air charts to build a three-dimensional picture of the atmosphere and its evolution in time. I also learned that the simple forecasting method of extrapolating past movements and trends into the future could be improved tremendously by considering the forces acting on weather systems and how those forces would be changing in the future. Perhaps most importantly, I learned about new tools that were coming into use, specifically satellites and computer models of the atmosphere. Satellite observations at that time were transmitted to local forecast offices via the weather facsimile machine. They were limited to poor-quality images of clouds associated with weather systems, but the potential for this new tool to fill in gaps in our observations and advance our knowledge of the atmosphere was already there. The new computer models were very simple

2. An excellent description of the New York City office of the U.S. Weather Bureau at that time is given in the April 1961 issue of *Weatherwise* magazine on pages 43–49 (Knudsen, 1961).

compared to the models of today, but these simple models were already proving to be very effective forecast tools. The primary output from these models was 500 mb heights and absolute vorticity over North America.

Vorticity, which is a mathematical measure of rotational motion, was a difficult concept for a 13-year-old to grasp. The forecasters defined it for me as cyclonic spin, but that was not enough for me to understand it. Fortunately, there were several publications there that I could read. One excellent pamphlet was entitled "Synoptic Meteorology as Practiced by the National Meteorological Center" (NMC). After reading this several times, as well as some Weather Bureau training manuals, I finally came to a sufficient understanding to use these charts effectively. This involved understanding the models and their limitations, modifying the models to correct for systematic errors, and determining the relationships between surface pressure systems and the upper-level vorticity field.

The 500 mb height and vorticity fields from these models had already become one of the primary forecast tools at the National Meteorological Center,[3] and their influence at local forecast offices was increasing rapidly. In preparing forecasts at this time, I would always begin with the local observations. Then I would make a detailed study of the surface weather map and prepare a candidate prognosis of the surface features based primarily on extrapolation of past movement and trends. Next, I would use the upper-air charts to modify the extrapolation of surface features and to further define the air mass characteristics and the causes of the weather being observed. Then I would begin my analysis of the output from the numerical models. This basic procedure of starting with local observations, then using surface and upper-air weather maps, and then utilizing the numerical models followed the historical evolution of forecasting methods.[4] I continued to use this approach, but with numerous additions and with greater emphasis on the numerical models as my career evolved.

From Mr. Fayne and the other forecasters I also learned about the American Meteorological Society (AMS). In 1963, I became an associate member of the AMS. I attended meetings of the New York City chapter of the AMS and one AMS annual meeting that was held in New York City. Most importantly, I received the *Bulletin of the American Meteorological Society* and *Weatherwise* magazine. I also ordered and read the American Meteorological Society's *Compendium of Meteorology*,[5] which covered a wide variety of topics in atmospheric

3. National Weather Analysis Center/Analysis and Forecast Branch, 1961.
4. Godske et al., 1957.
5. Malone, 1951.

science. All of these publications broadened my view of meteorology and increased my motivation even more.

By the time I graduated from high school and went off to college, I had developed a love of weather forecasting and some skill. For my undergraduate education, I chose Parks College of Aeronautical Technology of Saint Louis University. This was primarily because it offered 53 credits of meteorology courses, which was more than any other college that I could find. It also had an Air Force Reserve Officer Training Corps (ROTC) detachment, from which I was able to receive a scholarship that covered all of my tuition. The meteorology courses included general, synoptic, dynamic, and physical meteorology, meteorological instrumentation, hydrology, and climatology. In addition to these, there were courses in aeronautics, mechanics, engineering, thermodynamics, atomic and nuclear physics, and advanced mathematics, as well as all of the typical liberal arts courses that comprise an undergraduate education. All of the meteorology courses were taught by Professor Benjamin F. Abell. Fortunately, he was a truly outstanding teacher and I benefited from every one of the courses I had with him. Parks required a bachelor's thesis. For this, I did a research project on the use of vorticity in weather forecasting. This resulted in a 75-page paper in which I documented how forecasters should use vorticity charts and investigated the relative value of vorticity prognoses from two different numerical models. Surprisingly, I found that the most accurate predictions came from forecaster modification of the simpler of the two models. This was due to the ease with which forecasters could modify this model to correct for its limitations. This is also why this model continued to be operational for many years, even though increasingly advanced models were being developed and implemented.

Chapter 2

Weather Officer in the U.S. Air Force

In July 1970, I graduated from Parks College and was immediately commissioned as a weather officer in the United States Air Force. One month later, I reported to England Air Force Base in Alexandria, Louisiana to begin my tour of active duty. Upon arrival at the base weather station, I was assigned to a tech sergeant with a lot of forecasting experience who would train me as an aviation forecaster. I found out very quickly that forecasts for aviation are very different from the typical forecasts for the general public. Instead of predicting general conditions for sky cover, wind velocity, precipitation, and the high and low temperatures, it was necessary to predict the types, heights, and amounts of clouds, wind velocity, the visibility, the runway temperatures, how these would change on an hourly basis, the times precipitation would begin and end, and any hazards such as turbulence, icing, or thunderstorms that incoming or outgoing aircraft might encounter along their flight path. In addition to putting out the forecasts for our area on a very tight schedule and issuing watches and warnings for severe weather, the forecaster on duty supervised the local weather observer, operated the radar, analyzed local weather maps, and briefed pilots in the weather station, as well as on closed-circuit television, telephones, and on air to ground radio.

I completed my training very quickly and immediately began shift work as an Air Force weather officer. The work was challenging and placed tremendous responsibility on the forecaster, but it was exactly what I was hoping for.

Professor Abell had recommended the Air Force's Air Weather Service as one of the best places to start a meteorology career, and he was right. Daytime shifts were very busy with numerous pilots in the weather station, each requiring a briefing for the weather en route to their destination. At the same time, multiple phones could be ringing and the high-priority air to ground radio could be calling. All of this was sometimes happening right at the time that the forecast for our terminal needed to be issued and transmitted. To deal with this, I learned to arrive at the weather station an hour before my shift was to begin. I would use the time to analyze the situation and prepare my own prognostic charts (forecast maps) and a candidate forecast. Other forecasters asked me why I did this when centrally prepared prognostic charts were available on our facsimile machine. I explained that not only did this afford me the possibility of improving upon the centralized forecast guidance, but it also enabled me to determine how much confidence to place in a specific forecast. This and all of the forecasting methods that I had studied paid off in a big way in that this knowledge enabled me to build a strong sense of trust with the pilots and command staff at our base.

While in the Air Force, I had many interesting and very important forecast situations to deal with. One time a pilot told me he had a medical evacuation to make where the patient would most likely die if the flight could not be made. However, he also said that if they hit moderate or severe turbulence, then the patient might die en route. I checked the centralized forecast guidance and moderate turbulence was indicated. I then went through the turbulence forecasting methodology that I had learned and was able to make a forecast for only light turbulence. As a result, the pilot made a successful evacuation.

Another important forecasting situation involved a strong hurricane in the Gulf of Mexico that was moving toward Louisiana. Here I utilized existing subjective forecasting techniques, based primarily on upper-level steering patterns, and predicted that the storm would pass well to the south of our base. I forecast winds in central Louisiana to not exceed 35 mph and this could only occur in thunderstorms that might be embedded in the outer rain bands of the hurricane. The National Weather Service forecaster in Alexandria, Louisiana called me on the phone and asked what I was predicting. He evidently agreed with me because a little while later I heard my forecast on the radio being given to the public. However, people above me in the military were not as confident in my forecast. I received calls from various levels of Air Weather Service, and then from the Pentagon, all asking me to increase the severity of the weather in my forecast. Despite all of this pressure, I was confident in the forecast that

I had made and did not change it. The hurricane followed a path very similar to what I had predicted and winds did not go above 35 mph at our base. On another occasion, a 24–36-hour forecast that I made for a family outbreak of tornadoes to affect northeast Louisiana and Mississippi unfortunately verified. This was a case in which a rapidly intensifying cyclone associated with strong positive vorticity advection aloft (strong upper-level forcing for surface pressure falls and upward vertical motion) was moving into an area of high instability. This forecast was given to a pilot who had planned to fly into the area of most likely tornadic development at that time.

My most rewarding assignment came toward the end of my tour of active duty in the Air Force. This was to command a combat weather team during special training exercises being held near Myrtle Beach in South Carolina. During the two weeks of this exercise, I made forecasts, gave formal briefings to the senior staff, and interacted closely with the enlisted personnel on the team. Two colonels, one Army and one Air Force, were in charge of the exercises. They were both very tough and demanding. Fortunately, every one of the forecasts I gave them over the two weeks of the exercises verified completely. After I returned to England Air Force Base, I decided to make a trip to Parks College to serve as a guest lecturer talking about my experiences in the Air Force. I found that I enjoyed teaching almost as much as I enjoyed forecasting. This led me to consider going to graduate school, and I applied to New York University. But before my tour of active duty ended, Air Weather Service headquarters sent another colonel to interview me. He asked me why I was able to have a very high degree of forecast accuracy. I told him that it was because I had studied and knew a lot of forecasting methods, but years later, I realized that I had given an incomplete answer. I believe now that what makes a good forecaster is having a love of the weather and weather forecasting, and being able to pour oneself into the weather situation. This cannot be done in an optimal way if it is a part-time activity.

Overall, my forecast accuracy in the Air Force was approximately 95%, and I had several entire months with 100% forecast accuracy. However, I should mention that like all forecasters, not every forecast I made at England Air Force Base verified. A few were very humbling experiences. One in particular was a case in which we had zero ceiling (height of the lowest cloud base) and zero visibility in very dense fog. A pilot came into the weather station at about 7 a.m. that morning and asked when we would have VFR (visual flight rules) conditions of 3,000-foot ceiling and three-mile visibility. I told him that I expected it to improve to those conditions by 9:30 a.m. When he came back, the ceiling had lifted to 3,000 feet as predicted, but the visibility had only improved to two and a half miles. Three hours later, we were still at two-and-a-half-mile visibility, and he was furious. This occurred before high temporal and spatial resolution imagery from geostationary satellites was available. The standard forecast procedure at the time was to estimate the amount of radiational heating at the top of the fog layer using thermodynamic (temperature and moisture vs. height) diagrams, but this was far from precise. Much better forecast methods utilizing geostationary satellite imagery and vertical pointing radar exist today. One other very memorable forecast bust occurred when a fire formed near our runway. The smoke from the fire then drifted over the runway at verification time and lowered our visibility to much lower values than I had predicted.

Chapter 3

Graduate Study at New York University

After honorable separation from active duty in the Air Force, I remained in the inactive reserve for Air Weather Service for several more years, while at the same time beginning my graduate studies at New York University (NYU) School of Engineering and Science. I initially found it very hard to be a student again and almost flunked out in my first semester. Fortunately, this changed and I was able to once again become a good student. I graduated from NYU with a master of science (MS) degree in meteorology in two semesters and took and passed the qualifying exams in physics, mathematics, and foreign language to be accepted to the PhD program there. My research for the MS degree dealt with the atmosphere of Mars, as I have long been interested in the dust storms that occur on that planet on a variety of scales.

In the summer before my PhD coursework began, I was one of four graduate students selected to work as an intern for the Facilities Laboratory of the National Center for Atmospheric Research (NCAR) in Boulder, Colorado. The director of the Facilities Lab was Dr. David Atlas. Even though he was not related to me, I was very aware of his stature in our field. He was world-renowned as an outstanding scientist and was one of the pioneers of radar meteorology. At the Facilities Lab, I was initially assigned a physics problem associated with the displacement of balloons in the atmosphere. Later, I worked in the field on the National Hail Research Experiment. This was an enjoyable break from being behind a desk. I also got to interact with numerical modelers at NCAR working on new models for improved numerical weather prediction. This was an extremely valuable learning experience for me.

At the end of the summer in 1973, I returned to NYU to begin work on a PhD. Professor Jerome Spar, who taught the geophysical hydrodynamics and numerical weather prediction courses at NYU, hired me as a graduate research assistant on a National Aeronautics and Space Administration (NASA) grant, and also became my advisor. Professor Spar was both a tremendous research scientist and teacher. He presented complex, highly technical material in a logical way that was easy to learn. His research grant with NASA involved atmospheric response to sea surface temperature variations.[1] To support this grant I worked at NASA's Goddard Institute for Space Studies (GISS) while at the same time completing the coursework that was required. In addition to numerical weather prediction, I took courses in advanced dynamics, oceanography, geophysical modeling, air pollution, physics and chemistry of Earth's upper atmosphere, and planetary atmospheres. During this time, I also had my first college teaching experience, teaching physics at the State University of New York's Agricultural and Technical College in Farmingdale, New York.

In November 1973, I met Shirley, the woman who was to become my wife. Two weeks later we were engaged, and we were married six months later at the same time that I was taking my oral qualifying exams. The oral exams covered all of meteorology, oceanography, and geophysics, and they were really tough. There were four separate exams, each lasting three hours. I was very nervous and did very poorly on my first exam. Fortunately, I did much better on all of the remaining exams and managed to pass. It was now time to choose a PhD research topic. After talking with Dr. Jim Miller (then a postdoctoral research associate at GISS), I decided to develop an advective mixed-layer ocean model (an upper-ocean model that includes both horizontal transport and changes in the depth of the uppermost layer of the ocean, which is typically well mixed and of nearly uniform temperature) and conduct experiments on the predictability of sea surface temperature variations and how sea surface temperature anomalies form. With advice from both Jim and Professor Spar, and with the availability of NASA computing resources, I was able to complete my thesis research successfully in just eight months.

During much of my time as a graduate student working at GISS, I was the only synoptic meteorologist (a meteorologist specializing in weather analysis and forecasting) there. As such, I was often called upon to evaluate numerical

1. Spar and Atlas, 1975; Spar et al., 1976.

prediction experiments and to provide advice regarding the potential for satellite data to improve weather forecasting. I also started a Forecasting Section at GISS and made experimental forecasts and gave briefings every day that I was there. One of the goals of the Forecasting Section was to determine how far into the future accurate forecasts could be made. In some cases accurate forecasts could be made 7 to 10 days into the future, and in others even the next day's weather was very uncertain. This demonstrates the well-known observation that some weather situations are more predictable than others. Our goal here was to determine the reasons for the different ranges of predictability.

In late 1974, a request came to GISS to provide someone to give a one-hour lesson on weather to the anchors of a new show that the ABC television network was creating. I was asked to do this and readily accepted the daunting challenge of distilling the essence of weather and weather forecasting into a single one-hour session. This was the *AM America* show, which was the predecessor to *Good Morning America*. The lesson that I prepared concentrated on the development and movement of pressure systems and fronts and the weather associated with them. After giving the lesson, I was asked if I would like to do the weather forecasting for the show. I agreed and prepared national forecasts for the 10 months that this show was on the air. I would get up at 3:30 a.m., travel to Manhattan to the National Weather Service Forecast Office, prepare forecast maps for the nation, primarily by modifying the prognoses from available numerical models, and then travel to the ABC studio to brief the news staff. From there I would travel to GISS to complete my ocean modeling experiments for my doctoral dissertation. My thesis research quantified the role of horizontal advection in an ocean mixed-layer model, improved the ocean model's predictions of sea surface temperature variations, and also showed the importance of having accurate observations of surface wind velocity over the ocean.[2] This latter aspect became a major research activity for me (described in the next chapter). I presented my dissertation defense at GISS in February 1976 and was very pleased that many of the outstanding scientists there were able to attend.

2. Atlas, 1975.

Chapter 4

Research Scientist at NASA GISS and GSFC

Impact of Satellite Temperature Sounding Data

Shortly after receiving my PhD in meteorology and oceanography, I became a National Research Council resident research associate at NASA GISS. Although I was hired to continue my work on ocean modeling and atmosphere-ocean model coupling, I was soon moved to a new research activity. NASA and the National Oceanic and Atmospheric Administration (NOAA) were conducting joint experiments to assess the impact of satellite-derived temperature sounding data on numerical weather prediction. These experiments were known as the Data System Tests (DST), and they were a prelude to the Global Weather Experiment that was planned for 1979. Nearly all of the meteorology work at GISS was being reorganized to support the Data Systems Test that was planned for 1976 (known as DST-6). The goal of DST-6 was to evaluate the impact of NOAA-4 and Nimbus-6 polar-orbiting satellite data collected during the period from January to March 1976. This project was being led by Dr. Milt Halem, an outstanding applied mathematician and leader, from whom I learned a great deal.

Milt formed four groups to support DST-6. The first was a data assimilation group led by Dr. Michael Ghil. The objective of this group was to develop and test different data assimilation methodologies to make optimum use of the satellite temperature sounding data. Michael, a brilliant theoretician, began his work in a way that I didn't expect, but greatly appreciated. He arranged to meet with me by our weather map display and asked me how synoptic meteorologists like myself analyze the weather and do quality control of observations. Following this, he led the development, implementation, and testing of several different data assimilation methods. The last of these, the time-continuous statistical assimilation method (SAM), was a major advance in the utilization of satellite temperature sounding data. In this method, the satellite temperature soundings were assimilated at every model time step, rather than every six hours, which was the common practice at that time and for many years after at most operational weather prediction centers. A second group on atmospheric modeling was led by Dr. William J. Quirk. Here the goal was to improve the skill of the GISS general circulation model (GCM), primarily by increasing its resolution. In addition to the application of this model to numerical weather prediction, the model was later used in a variety of very innovative climate studies. A third group, led by Dr. Joel Susskind, sought to improve the accuracy and coverage of satellite temperature soundings. This group developed and implemented physical retrieval schemes, which over the years have benefited both weather and climate research. The last group was the meteorological evaluation group, which I was asked to lead. Initially, I utilized standard statistical scores and subjective evaluations of analyses and forecast maps to evaluate the impact of satellite data, but Milt wanted more. He first asked me to conduct an experiment in which forecasters (including myself) would make weather forecasts from model-generated maps with and without satellite data. To eliminate any bias that might exist in this experiment, the forecasters didn't know whether the charts they were looking at had included satellite data.

While this study was very useful, we soon found that forecasts could only be generated for a few cities because of the time required for each. Milt then asked me to create a computer code to simulate the way human forecasters use model guidance in making a forecast. For this purpose, I developed an expert system called the Automated Forecasting Method (AFM). The AFM was based on subjective forecast experience. It included various aspects of large-scale forcing, as well as objective determination of cold and warm fronts and their physical influences. The initial version of the AFM produced forecasts of precipitation occurrence and type of precipitation for 128 cities. Using only output from the GISS model, the AFM was found to be more accurate than the model's own precipitation forecasts. It was also found to be in excellent agreement with experienced forecasters using the same model data as input to their forecasts.[1] The AFM was not designed to replace human forecasters, who

1. Atlas et al., 1981.

use much more than the output from a single numerical model to produce their forecasts. However, it proved to be an outstanding verification tool for use in data impact studies. With the assistance of several students at the State University of New York at Stony Brook, I developed a second expert system, referred to as the Computerized Severe Storm Model (CSSM), at a later time. This system yielded predictions of high, moderate, or low potential for severe local storms.[2]

The initial experiments for DST-6 were conducted using the coarse resolution (four-degree latitude by five-degree longitude) version of the GISS GCM. These experiments tested each of the different data assimilation methodologies, and the assimilation of each satellite individually or both together. The evaluation of the impact that I performed included the magnitudes and locations of initial state differences for each of the forecasts, standard statistical measures of forecast accuracy, subjective comparisons of prognostic charts with and without satellite data, and verifications of local precipitation forecasts at 128 cities generated by the AFM with and without satellite data. By far, the largest, most beneficial impact of the satellite temperature soundings was found using the time-continuous statistical assimilation method to assimilate the data from NOAA-4 and Nimbus-6 in combination. The evaluation of initial states showed large spatially coherent differences in data-sparse regions. The statistical evaluation showed statistically significant improvements to forecasts in the 48- to 72-hour time range. The subjective evaluation that I performed and the AFM forecasts confirmed that the beneficial impact of the satellite data was meteorologically significant as well. These results were published as a NASA Technical Memorandum in January 1978, and in the American Meteorological Society's *Monthly Weather Review* in February 1979.[3]

What I didn't know at the time we were completing the evaluation was how controversial the DST-6 experiments would be. While we were running our data assimilation and forecasting experiments at GISS, NOAA's National Meteorological Center was running somewhat similar experiments using their operational data assimilation system. These were being led by Dr. Steven Tracton, whom I knew to be an excellent synoptic meteorologist. In contrast to the NASA DST-6 results, NMC found the impact of the satellite data to be small and of inconsistent sign. This averaged out to being only very slightly positive, and they concluded that the satellite temperature soundings had little impact

2. Atlas, 1978.
3. Halem et al., 1978; Ghil et al., 1979.

in the Northern Hemisphere.[4] Some of the meteorologists at NMC argued that their analyses and forecasts were more accurate than those generated at GISS, in part because of their model's higher resolution, and therefore were less likely to be improved by the assimilation of satellite temperature soundings. In addition, there was some speculation that as models went to higher resolution, the impact would become negative.

In contrast to this, we argued that NMC showed less impact because they did not assimilate the satellite data as effectively as we had done using the time-continuous statistical assimilation method. We then decided to rerun a limited portion of the DST-6 experiment with a much higher resolution version of the GISS GCM. In running this experiment, I found that, as predicted by NMC, doubling the model's resolution had a large impact on our forecast accuracy, and that the high-resolution forecasts without satellite data were better than the coarse resolution forecasts with satellite data. However, in the higher resolution experiment, the impact of the satellite temperature sounding data not only remained positive, but was larger and more meteorologically significant than it had been in our earlier experiment. The subjective evaluation that I performed was in excellent agreement with the statistical measures. No cases of significant negative impact occurred, while two very significant cases of major improvement, the forecasts from February 19 and 11, 1976, as well as other cases of less significant improvements were observed. Of these, the February 19 case was particularly important because of the severity of the weather that occurred. (This case will be discussed further in the section on major weather events and also in the chapter on teaching.)

I next compared the GISS high-resolution forecasts, with satellite data included, to NMC's forecasts and found that these forecasts were major improvements over NMC's with and without satellite data. I then carried out a detailed investigation of how the prognostic differences in these forecasts developed. Here I was able to trace the forecast impact at three days back to specific initial state differences due to the assimilation of satellite data and show how these differences grew with time as the forecast evolved.

Milt Halem decided that I should present these results to NMC before presenting them at the AMS Numerical Weather Prediction Conference in 1979. The one-hour seminar that I was scheduled to give lasted three hours because of all of the questions I received, but it had a beneficial result. At the conference, after I presented our DST-6 forecast impact results, two outstanding scientists

4. Tracton et al., 1980.

from NMC, Drs. Norm Phillips and Ron McPherson, went to the microphone to acknowledge the beneficial impact of satellite temperature soundings on numerical weather prediction and the desire to get this data into operations. I published the higher resolution impact results in a conference article and a NASA Technical Memorandum in 1979, and in two related *Monthly Weather Review* articles in 1982.[5] In 1981, scientists from NMC published a note in the *Monthly Weather Review*[6] entitled "On the System Dependency of Satellite Sounding Impact—Comments on Recent Impact Test Results." In this note, they stated: "An improved version of the GLAS system has been applied recently to demonstrate two cases (11 and 19 February 1976) for which positive impact was obtained and the GLAS system outperformed that of NMC. We are delighted that two cases of undisputed positive impact have been identified."

During the period in which we performed all of our DST-6 experiments, a major organizational change occurred. In 1977, NASA moved the global weather activity at GISS in New York to NASA's Goddard Space Flight Center (GSFC) in Maryland to become part of the Goddard Laboratory for Atmospheric Sciences (GLAS), which was being formed by Dave Atlas. I didn't move with the group initially, but continued to do research with them, while also teaching courses at several universities in New York. One day a week, I would travel to the Goddard Space Flight Center and bring back maps and data to analyze and evaluate. The ping-pong table in my basement was constantly covered with computer output that I was working with. After commuting back and forth between Maryland and New York for one year, I decided to join GSFC full time as a civil servant. In August 1978, I joined the Global Modeling and Simulation Branch (GMSB), which was headed by Dr. Halem, as a GS13 civil servant.

Goddard was an almost magical place to work: great leadership in Dave Atlas and Milt Halem, great colleagues like Drs. Eugenia Kalnay, Jagadish Shukla, Joel Susskind, Bill Lau, Louis Uccellini, Dave Randall, Ross Hoffman, Yogesh Sud, and many others. There were also almost endless opportunities for research into many different areas. Initially, our research focused on preparing for and then carrying out NASA's portion of the 1979 Global Weather Experiment.[7] My own work evolved into several different areas. I continued to work with satellite temperature sounding data, but also became heavily involved with improving and then optimizing the impact of satellite scatterometer ocean surface wind data.

---

5. Atlas, 1979, 1982; Atlas et al., 1979, 1982.
6. Tracton et al., 1981.
7. Halem et al., 1982.

I also became very heavily involved with observing system simulation experiments (OSSEs), Doppler wind lidar, and numerical diagnostic studies of major weather events. Throughout this period, I was able to interact with some of the top meteorologists in the world who often visited the Global Modeling and Simulation Branch. These included Professors Jule Charney, Ed Lorenz, Yale Mintz, Richard Lindzen, Donald R. Johnson, Fred Sanders, Michael Ghil, Jim O'Brien, and many others.

My later work with satellite sounding data was focused on the assimilation of advanced hyperspectral (high vertical resolution) sounding data from the Atmospheric Infrared Sounder (AIRS). Dr. Mous Chahine from NASA's Jet Propulsion Laboratory (JPL) was the leader of NASA's AIRS Science Team. At Goddard, Dr. Joel Susskind was a member of the AIRS Science Team and was leading the research to retrieve highly accurate weather and climate variables from AIRS. My job was to conduct research to assess the impact of AIRS temperature soundings on numerical weather prediction. I began this research well before the launch of AIRS by performing observing system simulation experiments (described later in this chapter). The purpose of these experiments was to determine the most optimum ways of assimilating the temperature sounding data that AIRS would provide. Following the launch of AIRS, I performed an extensive series of data impact experiments. These resulted in the first demonstration of improved numerical forecasts resulting from the assimilation of AIRS data. The magnitude of the impact agreed very well with what the OSSEs had predicted. Also in agreement with the OSSEs, the most beneficial impact occurred when AIRS temperature soundings retrieved under both clear and partially cloudy conditions were assimilated. Thanks to the tremendous work of Mous Chahine, Tom Pagano, and the entire AIRS Science Team and project, this was just the first of many significant advances resulting from the AIRS mission. The initial impact of AIRS data on numerical weather prediction and its potential for measuring greenhouse gases was published in conference proceedings in 2004 and in the *Bulletin of the American Meteorological Society* in 2006.[8]

I continued my interactions with the AIRS team and with JPL for many years after leaving NASA. In this regard, I attended and gave presentations at AIRS Science Team meetings, collaborated with AIRS Science Team members and other JPL scientists, and participated with them on several important research activities that were of interest to both NASA and NOAA. In 2010, NASA presented an

8. Chahine et al., 2006; Susskind and Atlas, 2004.

award to the NOAA AIRS Science Team, and I was very pleased to accept this award on behalf of all of the NOAA employees who worked with AIRS data. I returned to JPL in 2011 to present Dr. Chahine with an award from the American Meteorological Society's Integrated Observing and Assimilation Conference that I chair each year. His presentation on the use of AIRS to monitor and understand the spatial distribution of carbon dioxide in the atmosphere was chosen as the best in the conference. Unfortunately, Dr. Chahine passed away not long afterward, but I and many scientists will always treasure our interactions with him and the inspiration that he provided to virtually all who knew him. The JPL Center director invited me to speak at a memorial event being held in honor of Dr. Chahine. I was very pleased to represent NOAA at this event and to speak about Dr. Chahine's leadership of the AIRS team and the impact of advanced sounders like AIRS on numerical weather prediction.

Impact of Ocean Surface Winds

My work with scatterometry and with OSSEs began at the same time. An earlier OSSE had predicted that when the Seasat scatterometer (a radar capable of measuring returns of energy from the ocean surface) was launched, it would have a substantial positive impact on numerical weather forecasts. However, when Dr. Wayman Baker and I evaluated the actual impact of Seasat ocean surface wind data after its launch, we found the impact to be nearly negligible on average. To investigate this, I looked at the assumptions that were made in the OSSE and how this related to the real-world situation. The OSSE had assumed that Seasat provided a unique wind direction and speed, while the real Seasat data had substantial directional ambiguity. More importantly, the OSSE treated Seasat data as if it occurred well above the surface instead of near the surface of the ocean. Using this information, I determined that the impact of scatterometer data would be substantially better if the directional ambiguity could be reduced and if the effects of the surface wind observations could be accurately extended in the vertical.

To reduce the directional ambiguity, a new scatterometer with more antennas or a different type of antenna configuration would be needed. This, along with other studies, led to the development of advanced scatterometers such as the AMI on the European remote sensing satellite (ERS 1 and 2), the NASA scatterometer (NSCAT), and the SeaWinds scatterometer on Quikscat. To test the second point, I conducted experiments to evaluate how sensitive numerical model predictions are to errors in surface wind analyses, and then an OSSE to evaluate the potential improvements that could result from extending the surface wind analysis corrections in the vertical. I spent several weeks researching and investigating how to best extend the vertical influence of the surface wind data. This showed that "flow-dependent" vertical extension, where the influence of surface wind observations on higher levels depends upon the meteorological conditions, worked best. We then tested this vertical extension with real Seasat data for the famous storm that damaged the *Queen Elizabeth II* ocean liner. This resulted in a very significant improvement in the prediction of this storm and to my knowledge was the first demonstration that assimilation of scatterometer data could lead to significant forecast improvement. Later, we implemented flow-dependent vertical extension of satellite surface wind data in the GLAS global data assimilation system that our branch was running. This demonstrated the beneficial impacts of scatterometer data on global numerical weather prediction.

Shortly after the *QE II* storm demonstration, I served on NASA's Satellite Surface Stress Committee and was in charge of determining the meteorological requirements for NASA's next scatterometer (NSCAT). At about the same time, I was invited to a meeting at Joint Oceanographic Institutions in Washington, DC. While there I learned about the Defense Meteorological Satellite Program (DMSP) satellite that was to be launched in 1987. This satellite contained the special sensor microwave imager (SSMI). This instrument was to provide ocean surface wind speed data that could be useful in the years before the ERS and NSCAT scatterometers were to be launched. This proved to be very serendipitous. On the drive back to Goddard, I thought of several ways to assign directions to the SSMI wind speeds and assimilate them. (This was necessary since our data assimilation system was designed to assimilate wind vectors and not wind speeds.) Within days, I sent a proposal to the NASA headquarters Physical Oceanography Program to develop and evaluate six different methods for direction assignment and then generate ocean surface wind velocity fields for both meteorological and oceanographic use. The NASA Oceans program manager, Dr. Jim Richman, met with me at Goddard. He expressed some skepticism that what I was proposing would be successful, but in what I saw as a very smart move, he offered to fund the proposal for one year and then reevaluate it. We tested the methods to assign directions using both simulated data and Seasat ocean surface winds. Of the different methods, a variational analysis method (VAM) that was originally developed by Dr. Ross Hoffman for Seasat worked best. Together with my former student Joe Ardizzone, we began to adapt the VAM for SSMI and to apply it to the SSMI data that was being produced by Frank Wentz and his team at Remote Sensing Systems. One of my colleagues at Goddard estimated that this would take at least a year due to the extensive testing and reprogramming that it required. However, thanks to Joe's very innovative programming ability, we had it working in just three weeks. The global ocean surface wind velocity fields that we produced using it gained worldwide acceptance and became the most requested data set in NASA's Physical Oceanography Data Archive at that time. Several other important outcomes followed. The data impact studies that we conducted using both the NASA and NOAA global data assimilation systems demonstrated a beneficial impact, and these data were incorporated into NOAA's operational use. Second, this SSMI ocean surface wind data set evolved into the Cross Calibrated Multi-Platform (CCMP) data set, for which there have been more than two million downloads at the time of this writing. Together, these data sets were funded by NASA for more than 20 years and have been used all over the world in multiple scientific disciplines. In addition to their wide use in weather and climate research and in the study of ocean processes, these data have been used in monitoring the movements of debris and radiation in the oceans, and in studies of how whales congregate.

With the launch of each successive scatterometer developed by NASA, I served on the science teams for these instruments. This was both very exciting, as we discovered additional features in the atmosphere, and very enjoyable because of the great colleagues who were on the teams. They included Mike Frielich, Tim Liu, Ross Hoffman, Frank Wentz, Bill Pierson, Jim O'Brien, and Deborah Smith, to name a few. As each of the new scatterometers was launched, our group at Goddard, consisting of Joe Ardizzone, Joe Terry, Genia Brin, Juan-Carlos Jusem, Steve Bloom, Dennis Bungato, and me, optimized their use in our global data assimilation system and demonstrated their beneficial impact. As with our SSMI work, we also briefed NOAA modelers and forecasters on our results. This was particularly rewarding, as the forecasters at NOAA's Ocean Prediction Center became very skilled at using scatterometer data, and this helped to save lives, ships, and cargo at sea. Dr. Paul Chang and his group at NOAA NESDIS ensured that the data got to the forecasters in real time, and also provided additional training to the forecasters. Much of our work with ocean surface wind data was summarized in the American Geophysical Union's *Journal of Geophysical Research* in 1984, 1986, 1987, and 1999, and in the *Bulletin of the American Meteorological Society* and the *Monthly Weather Review* in 1993, 1996, 2001, and 2011.[9]

---

9. Atlas, 1996a,b; Atlas et al., 1987, 1996, 1999, 2001, 2011; Baker et al., 1984; Bloom et al., 1996; Duffy and Atlas, 1986; Duffy et al., 1984; Kalnay and Atlas, 1986; Lenzen et al., 1993.

Observing System Simulation Experiments

Observing system simulation experiments (OSSEs) are simulation experiments designed to assess the potential impact of new or proposed observing systems. They may also be used to evaluate trade-offs in instrument design or to develop, evaluate, or improve data assimilation methodology. OSSEs consist of the following elements: a nature run, which is a long integration of a numerical model (a forecast); simulated observations; a data assimilation system; and forecasts with and without the observing system being evaluated.

In the early 1980s, Milt Halem and Eugenia Kalnay asked me to take the lead on the OSSE work at Goddard. I began by looking at past OSSEs, especially where they had not given accurate quantitative results. I then surveyed several key scientists who were familiar with OSSEs about what they felt were the limitations and weak points. I still remember the conversation with Dr. Norm Phillips, who was at the time the chief scientist at NMC. He said the biggest problem was the nature runs, which weren't sufficiently realistic. Afterward, I set about modifying OSSE methodology to make these experiments more credible. This included using progressively more realistic global models to generate nature runs, simulating observations to better take account of their coverage and accuracy, using a sufficiently different model (from the nature run model) to assimilate the simulated observations and make forecasts, and most importantly, validating the OSSE system by comparisons with real data impact experiments. This modified OSSE methodology received both national and international acceptance. With the help of an outstanding support staff, and especially Genia Brin and my former student Joe Terry, I completed my first OSSE in 1985 and then performed all of NASA's global weather prediction OSSEs from 1985 to 2005. These OSSEs proved to be especially valuable in making decisions regarding proposed observing systems and in developing improved data assimilation methodologies.

In addition to the OSSEs to improve the assimilation and impact of satellite surface wind data from scatterometers and from SSMI, we conducted extensive global weather prediction OSSEs for a wide variety of purposes. One of the early OSSEs I performed evaluated the relative importance of temperature, wind, and moisture observations. These experiments showed wind profiles to be much more effective than temperature or moisture profiles in correcting analysis errors and indicated significant potential for space-based wind profiles to improve numerical weather prediction. A related OSSE that followed evaluated the relative importance of upper- and lower-level wind data. This experiment showed that wind profile data from the 500 mb level and higher in the atmosphere provide most of the impact on large-scale numerical forecasting.

Additional OSSEs evaluated trade-offs relating to coverage and accuracy for satellite observing systems and were used to define draft requirements for lidars[10] (laser radars) in space. The quantitative and relative impacts of NSCAT and ERS that were predicted by our OSSEs agreed almost exactly with the actual impacts that were found after their launch. Similarly, the impact of the advanced Atmospheric Infrared Sounder (AIRS) and the effect of clearing clouds from the AIRS data also verified extremely well after the launch of AIRS. To my knowledge, these were the first instances when the quantitative impact predicted by OSSEs verified after the launch into space of the satellite observing systems that had been investigated. This added great credibility to OSSEs. The results of all of the OSSEs that I performed at NASA and later at NOAA were published over a period of many years in refereed articles and conference proceedings and included many co-authors.[11] In addition to Joe Terry and several others at Goddard who were part of the experiments, Dr. Dave Emmitt of Simpson Weather Associates and members of his staff became trusted partners in all of the lidar wind experiments that we performed. Similarly, Drs. Ross Hoffman and Lars Peter Riishojgaard, and Professor Zhaoxia Pu also became trusted partners in a number of the OSSEs that I have been involved in.

Studies of Major Weather Events

In addition to my research to enhance the utilization of satellite data, I devoted a considerable amount of time at GISS and GSFC researching a variety of significant weather events. These included extratropical and tropical cyclones, severe local storms, and prolonged heat waves and drought.

The first storm I studied in great detail was from one of the cases of improved prediction that resulted from the assimilation of DST-6 satellite temperature sounding data. On February 19, 1976, a weak low pressure system was located off the west coast of the United States. This storm underwent a center jump (an apparent jump of a low pressure center to a new location, usually slightly further downstream) due to the development of a new low pressure

10. Baker et al., 1995, 2014.

11. Annane et al., 2017; Atlas et al., 1984, 1985a,b, 2015a,b; Atlas, 1997; Atlas and Emmitt, 1991, 2008; Atlas and Riishojgaard, 2008; Atlas and Pagano, 2014; Prive et al., 2014; Leidner et al., 2017; McNoldy et al., 2017; Pu et al., 2017; Zhang et al., 2017; Cucurull et al., 2018; Mueller et al., 2020.

center within the original low's circulation, moved southeastward across the Rocky Mountains, and then recurved (changed direction) to the northeast. It produced blizzard conditions to its northwest, and as it recurved it underwent significant intensification and was associated with numerous severe thunderstorms and tornadoes. For this storm, I studied how it intensified and why it recurved in the real world and in the forecast that included DST-6 satellite data, but not in the forecast without this data. I also applied the CSSM expert system for severe local storm prediction to this case, which my students at the State University of New York and I had developed. The goal here was to understand the mechanisms for the severe weather development and why the forecast with DST-6 satellite data accounted for much more (70% vs. 40% for a 48–60-hour forecast) of the severe weather development. Most of the results from this study were published in the *Monthly Weather Review* in 1982 and 1984.[12]

Another cyclone that I studied in great detail was the Mid-Atlantic States Cyclone of February 18–19, 1979 (more commonly known as the President's Day Cyclone). Here, truly outstanding research on this storm was done by several other meteorologists, especially Drs. Louis Uccellini and Lance Bosart.[13] My research focused on the GLAS model simulations of this storm. Together with my former student Robert Rosenberg, I conducted a series of forecast experiments to assess the accuracy of the GLAS model's predictions of this storm and to determine the importance of large-scale dynamical processes and air-sea interaction to the cyclogenesis (development of a new low pressure system) that occurred.[14] The 36-hour GLAS model forecast from the GLAS analysis at 0000 UT 18 February correctly predicted strong cyclogenesis and heavy snow in the correct location, despite its coarse horizontal resolution. This forecast was repeated without surface heat or surface heat and moisture fluxes from the ocean to atmosphere, and then again with different initial conditions that had much weaker upper-level forcing for cyclone development. Examination of these forecasts showed that the heating resulting from oceanic fluxes contributed significantly to the generation of low-level cyclonic vorticity and the intensification and slow rate of movement of an upper-level (high pressure) ridge over the western North Atlantic. As an upper-level trough of low pressure approached this ridge, warming associated with the release of latent heat intensified, and the gradient of vorticity, vorticity advection, and upper-level

12. Atlas, 1982; Atlas et al., 1982; Vergin et al., 1984.
13. Bosart, 1981; Uccellini et al., 1984, 1987.
14. Atlas and Rosenberg, 1982; Atlas, 1987.

divergence in advance of the trough were increased very substantially. This provided strong large-scale forcing for the intensification of the cyclone and the very heavy snow (more than two feet) that occurred. We also studied this case from a forecaster's perspective. This will be discussed later in this volume.

My research on heat waves and drought began in the summer of 1980. At that time, the United States and especially the Great Plains was experiencing a severe heat wave. (Dallas, Texas had already experienced over 40 straight days of maximum temperatures exceeding 100 degrees F). One day, Milt Halem asked me why it was so hot. I answered that there was an abnormally strong upper-level ridge of high pressure over the central United States and that this ridge was increasing the heating and suppressing precipitation. He then asked why the ridge was so anomalously strong. To answer that required research. This began as a collaboration with Dr. Noah Wolfson (a visiting scientist from Israel), Dr. Yogesh Sud (a climate scientist and modeler in our branch), and Joe Terry (my former student from the University of Maryland who also worked with me on nearly all of the OSSEs I performed while at NASA).

Since this was a new area of research for me, I began it with a literature search, in which I reviewed many of the previous studies of heat waves as well as all of the scientific and newspaper accounts of this heat wave. We next developed and published a heat wave index that could be applied to model analyses and forecasts, and then conducted a series of numerical modeling experiments that showed the mechanisms involved in the development, maintenance, and breakdown of the upper-level ridge and the heat wave. We also studied the predictability of this heat wave and found strong dependence on sea surface temperatures in the North Pacific and soil moisture over the continental United States. This dependence on boundary conditions was expected, but we also found some dependence on the initial state. This latter dependence was unexpected given the length of the model forecasts that we generated. The GLAS model simulations from mid-May 1980 correctly indicated the development, maintenance, and breakdown of the heat wave, while comparable simulations from mid-May 1979 did not. This raised numerous questions about the predictability of prolonged heat waves and drought and led to more research by us and other scientists. The results of our experiments with the 1980 heat wave were published in the journal *Atmosphere-Ocean* and in several conference articles from 1985 to 1988, and in the *Monthly Weather Review* in 1986.[15]

15. Wolfson and Atlas, 1986; Wolfson et al., 1985, 1987; Atlas et al., 1988.

The second major heat wave that we studied while I was at NASA Goddard was the severe 1988 U.S. heat wave and drought. This heat wave caused approximately $30 billion in agricultural damage and contributed to the deaths of 10,000 people from heat stress. Here Noah Wolfson, Joe Terry, and I performed a series of numerical experiments aimed at quantifying the relative importance of tropical and mid-latitude sea surface temperature (SST) anomalies and soil moisture (SM) anomalies on the heat wave and drought over the Great Plains. In the *Journal of Climate* paper that we published in 1993,[16] we documented (I believe for the first time) the relative importance of each of these anomalies on the Goddard model prediction of surface temperature and precipitation. In addition to these results, we had one very unexpected finding. We observed that the soil moisture anomalies over North America not only had local and downstream effects but also had significant upstream effects as well.

Management at NASA GSFC

Over the 27 years that I was a civil servant at NASA's Goddard Space Flight Center, many management and organizational changes occurred. At the laboratory level, Dave Atlas was followed by Marvin Geller and the name of the lab was changed to Goddard Laboratory for Atmospheres (GLA). Marvin was followed by Franco Einaudi and then by Bill Lau as the laboratory chief. At the branch level, Milt Halem left to become the head of the computing division at GSFC. He was followed by Dr. Eugenia Kalnay, and then by Dr. Ray Bates after Eugenia moved to NOAA to lead their Environmental Modeling Center. After this, our branch was reorganized into two offices: the Data Assimilation Office (DAO) and the Satellite Data Utilization Office (SDUO). I chose to go into the SDUO, which was to be led by my good friend Dr. Joel Susskind. Within the SDUO, I headed the Synoptic Evaluation Group. Later this group and I were moved into the DAO, and I was asked to become the head of the DAO.

This was at a time when the DAO was under tremendous pressure. I struggled with the decision, as I had never aspired to be a manager and there were many obstacles to being successful. However, I had many friends in the DAO who were contractors, and I felt that many of them might lose their jobs if the DAO didn't succeed. So I decided to give it a go, with the hope that I could return to being a research scientist within a year or two.

16. Atlas et al., 1993.

I began this task by going to NASA headquarters and meeting with the chief scientist for Earth Sciences, and several key managers there, in order to find out what they felt were the weaknesses of the DAO. I was told that the budget for the DAO was in danger of being reduced substantially or even eliminated unless we could improve the DAO's global model (known as GEOS-1) and be ready for the launch of NASA's Earth Observing System (EOS). I was also told that I needed to attend the next meeting of the Scientific Steering Group (SSG) for the Global Energy and Water Cycle Experiment (GEWEX), and present what I would be doing to improve GEOS-1 and our products for the scientific community. They also added that I should expect severe criticism from the GEWEX SSG, but it was necessary that I go there. Afterward, I met with several of the key people in the DAO and very quickly prepared my plan for how to proceed. This began with immediately doubling the resolution of the current version of the GEOS model and forming a new model development group. This group was tasked with the development of a "next-generation" global model based on finite volume dynamics[17] and was to be led by Dr. Shian-Jiann (S-J) Lin. Substantial developments to the DAO data assimilation system were also planned, as well as major improvements in how the DAO functioned. There was substantial resistance to several of the changes that I was making, but I motivated the staff to embrace these changes and began implementing them. One additional change that I made was to change the name of the DAO from the Goddard Data Assimilation Office to the NASA Data Assimilation Office. This was designed to involve all of the NASA centers and their partners in the work of the DAO.

Tremendous progress, especially with regard to global modeling,[18] was made very quickly. At the GEWEX meeting, I presented my plan, the progress that had already been achieved, and a detailed evaluation of the model's performance. I showed verification of the model across the spectrum from climate to weather. One of the GEWEX SSG members stated that this was the most comprehensive validation of a global atmospheric model that he had ever seen. NASA HQ received very favorable reviews of what we were doing, and the GEWEX SSG voted to include me as one of their members. I served as the head of the DAO for five years and oversaw the development of the GEOS-2, 3, and 4

17. For a description of finite volume dynamics and their early application to numerical modeling see Lin (2004).

18. Conaty et al., 2001.

data assimilation systems. One of the best accomplishments of the DAO was the finite volume General Circulation Model (fvGCM), thanks to the fantastic skill and leadership of S-J Lin and his tremendous support staff. The dynamical core for this model was requested and used by organizations around the world, and it also provided the starting point for the research that led to NOAA's current operational model. Other accomplishments were the development of an advanced data assimilation system using the fvGCM by Dr. Arlindo da Silva, and numerous advances in knowledge, methodology, and utilization of new types of satellite data. Nevertheless, heading the DAO was very stressful and leadership of the DAO was eventually transferred to another scientist as the DAO merged with another organization. At that time the name was changed to the Global Modeling and Assimilation Office (GMAO), and this organization continues to perform exceptionally well.

In 2003, the chief of the Laboratory for Atmospheres, Dr. Bill Lau, asked me to become the chief meteorologist for the lab and I accepted. I was thrilled to be working directly with and for Bill, who was and still is an outstanding scientist. It was great to be back working as a research scientist again, but there was one thing I missed about being the head of an organization. I found that I really enjoyed being able to help advance other scientists' careers and not just my own. This and the fact that I had family in south Florida were both strong factors in my decision to apply to become director of NOAA's Atlantic Oceanographic and Meteorological Laboratory in Miami, Florida.

I served as chief meteorologist at GLA for two years. During this time, I served as the lead for hurricane modeling research, described in the next section, and served as principal investigator on 12 proposals and on NASA Science and Instrument teams. Also at the request of NASA headquarters, S-J Lin, who had left NASA in 2003 to join NOAA's Geophysical Fluid Dynamics Laboratory (GFDL), K. Yeh, and I prepared a 10-year plan for the development of a global cloud-resolving earth system model based on finite volume dynamics. This plan had informal buy-in from multiple agencies and universities, but unfortunately, a change in priorities resulted in fewer resources being available. Still, components of both NASA and NOAA have followed aspects of this plan, and great progress has been made in both agencies.

Hurricane Modeling with the fvGCM

After S-J Lin left NASA to join NOAA GFDL in 2003, I served as the head of the fvGCM modeling group at Goddard. During this time, I stayed in close contact with S-J, but we also had several outstanding modelers in the group. This included Drs. Bill Putman, Bo-Wen Shen, and Jiun-Dar Chern. In addition, Dr. Oreste Reale, who had been a student in my synoptic meteorology classes at the University of Maryland, worked very closely with the modeling group, especially with regard to the model's ability to simulate hurricanes and other tropical phenomena. Prior to the 2004 hurricane season, Bo-Wen had brought the fvGCM to approximately .25 degree (approximately 25 kilometer) resolution, making it one of the highest resolution global models to be run in real time in 2004. In 2005, Bo-Wen increased the resolution further, bringing it first to .125 degree and then to approximately .07 degree horizontal resolution.[19]

Our initial tropical cyclone experiments with the fvGCM consisted of 15 five-day simulations of four Atlantic tropical systems that occurred in 2002 and 2004. These storms were selected because of their complex tracks and their radically different life cycles. The model demonstrated the ability to simulate all of these different systems, which included abrupt recurvature, intense extratropical transitions, and multiple landfalls with reintensification and interaction among tropical cyclones. After these initial experiments, we ran the fvGCM in real time at both .5 degree and .25 degree resolutions for every Atlantic tropical system during the very active 2004 hurricane season. Early in this process, Oreste developed guidance for forecasters on how to best use the fvGCM model output. This guidance was designed to help forecasters know under what conditions the fvGCM would perform well and when it would not. I used this guidance throughout the 2004 hurricane season to make experimental hurricane forecasts and found that it was extremely effective in letting me know when to believe and when not to believe the fvGCM. This was especially important whenever the fvGCM was different from most or all of the other model forecasts. Many spectacular forecasts occurred in 2004 where the fvGCM was an outlier from other models. For example, the fvGCM five-day landfall forecast for Hurricane Ivan had an error of less than 50 kilometers, while other forecasts had landfall errors nearly ten times larger. In addition, the fvGCM produced a very good intensity forecast for Ivan, correctly predicting that it would make landfall as a category 3 hurricane. The five-day landfall forecast for Hurricane Jean was even better, missing the landfall location by only two kilometers and the timing of the landfall by only two hours. NASA Public Affairs asked me if we should put out a press release about how much NASA was improving hurricane forecasting. I replied that we would have to run the model for several seasons before we could have confidence in the results. When

19. Atlas et al., 2005, 2007; Shen et al., 2006a,b, 2010.

I presented the results with the fvGCM at the Interdepartmental Hurricane conference that year, I was very clear in indicating that while the initial results with the fvGCM were very encouraging, it was just a first step and a great deal of research still needed to be done. This led to my taking several trips to Miami and increasing interactions with scientists at NOAA's Atlantic Oceanographic and Meteorological Laboratory.

Plate 1. Air Force ROTC cadet Atlas preparing for a training flight aboard a T33 jet aircraft in 1969. (When the Air Force pilot turned the control of the aircraft over to me, I used the opportunity to investigate how turbulence affected small jet aircraft in the variety of clouds present that day.)

Plate 2. Lt. Robert Atlas in the weather station at England Air Force Base circa 1972.

Plate 3. Assistant professor Atlas making a forecast at the SUNY Stony Brook Weather Observatory in 1976.

Plate 4. Shirley Atlas (left), Bob Atlas (center), and Bob Rosenberg (right) at NASA Goddard Space Flight Center in 2005.

Plate 5. Left to right: Frank Marks (director of HRD), Max Mayfield (former director of NHC), Bob Atlas (director of AOML), and Bill Read (former director of NHC). Photo courtesy of NOAA AOML.

Plate 6. Left to right: Bob Atlas, VADM Conrad Lautenbacher (NOAA administrator 2001–2008), and David Atlas (former chief of the Goddard Laboratory for Atmospheric Sciences) at the 50th anniversary of the National Hurricane Research Laboratory hosted by AOML, May 22, 2006. Photo courtesy of NOAA AOML.

Plate 7. Left to right: Silvia Garzoli and Gustavo Goni (past directors of PHOD), Jane Lubchenco (NOAA administrator, 2009–2013), Molly Baringer (deputy director of AOML), Bob Atlas, and Craig McLean (director of OAR) on the occasion of Dr. Lubchenco's visit to AOML, October 20, 2011. Photo courtesy of NOAA AOML.

Plate 8. Left to right: Bob Atlas, Gary Matlock (deputy assistant administrator for OAR), RADM Tim Gallaudet (assistant secretary of commerce for oceans and atmosphere, 2017–2021), and Dr. Neil Jacobs (assistant secretary of commerce for environmental observation and prediction, 2018–2021) on the occasion of their visit to AOML, April 4, 2018. Photo courtesy of NOAA AOML.

Plate 9. Joe Ardizzone (left) with Bob Atlas at the AMS Annual Meeting in 2011.

Plate 10. Dr. Atlas, chair of the AMS Integrated Observing and Assimilation Systems for the Atmosphere, Oceans, and Land Surface Conference, presenting Dr. Mous Chahine with a certificate for having given the most outstanding presentation at the conference in 2011. Photo by Sharon Ray at NASA's Jet Propulsion Laboratory, March 2011.

Plate 11. Bob Atlas with Professor James R. Miller of Rutgers University on one of Jim's visits to AOML. Photo courtesy of NOAA AOML.

Plate 12. Congresswoman Debbie Wasserman Schultz and Bob Atlas briefing the news media on the upcoming hurricane season. Rep. Wasserman Schultz organized these briefings every year at the start of the hurricane season to ensure that the federal government and local governments were fully prepared, and also to encourage local residents to take the forecasts and warnings very seriously. Each year, I would present the seasonal outlook and the advances in hurricane research, and answer questions from the media about our models and observing systems. Most years, Ed Rappaport would represent the National Hurricane Center and report on advances in forecasting and new products that had become operational. Photo courtesy of NOAA AOML.

Plate 13. Judy Gray (left) and Bob Atlas (right) with Congressman Mario Diaz Ballart (center) at the May 12, 2007 AOML Open House. Representative Diaz Ballart was always very interested in all of the research being performed at AOML, and our employees and I very much appreciated his interest. Photo courtesy of NOAA AOML.

Plate 14. Bob Atlas presenting certificates of appreciation to noted hurricane researcher Dr. Pete Black upon his retirement from NOAA. Photo courtesy of NOAA AOML.

Plate 15. Bob Atlas recognizing Dr. Sundararaman Gopalakrishnan for his leadership of the hurricane modeling group at AOML HRD. Photo courtesy of NOAA AOML.

Plate 16. NOAA Corps officer Hector Casanova and Bob Atlas aboard the RV *Hildebrand*. The *Hildebrand* was one of AOML's small boats that was used primarily for coastal ocean research by AOML's Ocean Chemistry and Ecosystems Division. Photo courtesy of NOAA AOML.

Plate 17. Left to right: Bob Atlas, Lisa Bucci, and Professor David Nolan receiving the American Meteorological Society's Banner I. Miller Award from AMS president Fred Carr. Photo courtesy of the American Meteorological Society.

Plate 18. Ed Rappaport, deputy director of NHC, presenting an award to Bob Atlas for enduring contributions to the nation's hurricane forecast and warning program, February 21, 2019. Photo courtesy of NOAA AOML.

Plate 19. Bob Atlas speaking at the National Hurricane Center, February 21, 2019. This was the third time I spoke at NHC. The first two times were seminars from my research. In this much briefer talk, I thanked NHC's Director Ken Graham, Deputy Director Ed Rappaport, and all of their staff for the great job that they do and for the excellent relationship that developed between AOML/HRD and NHC. Photo courtesy of NOAA AOML.

Plate 20. Flyer and agenda for the Atlas Science Symposium that was hosted by the Cooperative Institute for Marine and Atmospheric Studies at the University of Miami's Rosenstiel School. All of the presentations are available online. The morning sessions can be viewed at https://youtu.be/oGabQGozCWI, while the afternoon sessions can be found at https://youtu.be/5fqPXTgiisE. Photo courtesy of NOAA AOML.

Plate 21. Attendees at the Atlas Science Symposium, February 26, 2019. Photo courtesy of NOAA AOML.

Plate 22. Old friends from NASA Goddard reunited at the Atlas Science Symposium. Left to right: Eugenia Kalnay, Bob Atlas, Milt Halem, Louis Uccellini, and Jagadish Shukla, February 26, 2019. Photo courtesy of NOAA AOML.

Plate 23. Michael Liquerman, press secretary/outreach coordinator for Congresswoman Debbie Wasserman Schultz, presenting Bob Atlas with the Congressional Record for February 19, 2019, at the reception that followed the Science Symposium. Photo courtesy of NOAA AOML.

Plate 24. RADM Nancy Hahn (deputy director of the NOAA Corps) presenting Bob Atlas with the NOAA flag that had flown at AOML for his arrival on August 22, 2005. Photo courtesy of NOAA AOML.

Plate 25. Ken Atlas (left) with Bob Atlas on a break from the Atlas Science Symposium. Photo courtesy of NOAA AOML.

Plate 26. Dave Jones (right) with Bob Atlas at the 2019 Tropical Cyclone Operations and Research Forum at the University of Miami, March 13, 2019. Photo courtesy of Heather Jones.

Chapter 5

Director of NOAA's Atlantic Oceanographic and Meteorological Laboratory

The Atlantic Oceanographic and Meteorological Laboratory (AOML) is one of NOAA's major research organizations. It consists of three divisions: the Physical Oceanography Division (PhOD), which does open ocean and climate research; the Ocean Chemistry and Ecosystems Division (OCED), which does coastal ocean and marine ecosystem research; and the Hurricane Research Division (HRD), which does research on tropical meteorology and, of course, tropical cyclones. The mission of AOML is to conduct research to understand the physical, chemical, and biological characteristics and processes of the ocean and the atmosphere, both separately and as a coupled system. The principal focus of these investigations is to advance knowledge that leads to more accurate forecasting of severe storms, better use and management of marine resources, better understanding of the factors affecting both climate and environmental quality, and improved ocean and weather services for the nation.

For several years while I was still at NASA GSFC, I had been interacting with Dr. Frank Marks, the director of the HRD. Frank is an outstanding hurricane scientist and I had involved him in our modeling and OSSE work as it related to hurricanes. On one of my visits to AOML, I met with the then-current director of the laboratory, Dr. Kristina Katsaros, who was planning to retire from NOAA. Kristina and I had worked together on the NASA Scatterometer Team for many years, and she encouraged me to apply for the position she would be vacating. I then met with AOML's deputy director, Judy Gray, and each of the division heads at AOML: Drs. Silvia Garzoli, Peter Ortner, and Frank Marks. I found all of the leadership at AOML to be highly competent and knowledgeable. They all encouraged me to apply for the director position, and I then began the exhaustive application process. After what seemed like a very long time, I was called for an interview, and not long after was offered the position by the director of NOAA's Office of Oceanic and Atmospheric Research (OAR), Dr. Rick Rosen. Dr. Rosen gave me some very specific directives before hiring me, though. He asked me to view my position as not just leading AOML, but working with all of NOAA to support NOAA's mission and to enhance its value to the nation. Dr. Rosen is someone I have great respect for and I took these directives very seriously. He was most concerned with the relationship of AOML to the Pacific Marine Environmental Laboratory (PMEL), which was the other lab within OAR that did significant open ocean research. He also encouraged me to work closely with the National Weather Service, especially with regard to our hurricane research activities.

On August 21, 2005, I became the director of AOML. On my first day there I held an all-hands meeting in which I listened to the ideas and concerns of the staff and briefed them on my philosophy of management and my plans for the laboratory. My way of leading didn't come from a book. It came primarily from experience. In the course of my career, I learned to emulate the positive things that I observed in other leaders. At the same time, I try very hard not to do any of the negative things that I observed. It's a very simple process, but one that I believe works very well. At the all-hands meeting, I expressed my basic philosophy that in the workplace everyone is deserving of respect and that we accomplish the most when we recognize our limitations and work with others who have complementary skills and knowledge. As such, my main goal for AOML's staff was to provide a creative and respectful workplace where all employees would enjoy coming to work and be able to achieve their maximum potential.

There were also several scientific goals that I brought with me to AOML. One was to bring numerical modeling back to the laboratory. Another goal was to create OSSE capabilities applicable to all of NOAA's missions. I also knew that within OAR, each of the laboratories was judged by the number and quality of its refereed publications and by the number and significance of its transitions of research to operations and applications. As such, another of my goals was to increase AOML's publications and transitions significantly. Finally, one of the major concerns that I heard from employees at the all-hands meeting was that AOML seemed to be largely invisible, with Congress and the public hardly knowing that we exist. With this concern in mind, I set an additional goal of correcting this, understanding that this would be an important factor in enhancing morale at the laboratory. To achieve all of these goals, I wanted to increase AOML's collaborations within NOAA, with other agencies, and with the academic community, and also to make sure that both Congress and the public were aware of the outstanding work being performed at AOML.

I followed the all-hands meeting with in-depth meetings with each of AOML's divisions. I found all of the research being performed at AOML to be worthwhile, but some important activities were missing. In the past, AOML had been very active in hurricane modeling, but none of the modelers were still there. Similarly, there was a strong need for ocean and ecosystem modeling to be performed at the laboratory.

My Involvement in AOML's Hurricane Research

During my first few months at AOML, my wife and I lived in hotels in Miami during the week, and we stayed with our son in the Florida Keys on the weekends. On my first weekend there, just prior to becoming director of AOML, I decided to do a quick survey of how people in the Keys would react to a severe hurricane threat. My wife and I walked through our son's neighborhood and asked all of the people we encountered if they would evacuate if a category 5 hurricane was forecast to hit them directly. Although this survey can't be viewed as scientific, every person we met said that they had been warned too much and they would not evacuate. In my opinion, the National Hurricane Center (NHC) was already doing an outstanding job, and the analysts and forecasters there were extremely knowledgeable, talented, and motivated. However, it was clear that there was a strong need to improve hurricane forecasts beyond the current "state of the art" so that public confidence in the warnings would increase and more lives and property would be saved. The 2004 hurricane season contributed to this awareness, but the 2005 hurricane season made it even more clear.

A few days after my arrival at AOML, Tropical Storm Katrina was located just to the east of Miami. It intensified to a category 1 hurricane just before making landfall and caused considerable damage in south Florida before moving into the Gulf of Mexico. For my family and me, this was a very exciting event and a learning experience. We were staying on a high floor of a Miami hotel when the eyewall of Katrina moved directly over us. Windows in the hotel were shattered, many trees came down, and electric power was lost throughout most of the county. In the Gulf, Katrina intensified rapidly to a category 5 hurricane and then weakened to category 3 just before landfall in southeastern Louisiana. The damage there and in Mississippi was catastrophic. The 2005 hurricane season continued to be extremely active and ended with 28 named storms, 15 hurricanes, and 7 major hurricanes. This included Hurricane Wilma, which underwent extremely rapid intensification and like Katrina caused substantial loss of life and property. With Wilma as with Katrina, I got to see what life is like after a hurricane if people and communities are not fully prepared. After both storms there was no electricity for long periods, gasoline for automobiles was almost nonexistent, and it was also very difficult to get food as restaurants and stores without generators were all closed. All of this motivated me and I'm sure every meteorologist at NHC and HRD even more. I thought that NOAA was very smart to have its main hurricane research activity in Miami.

Due in large part to the very active 2004 and 2005 hurricane seasons, and the recognition that improvements to hurricane forecasting (especially forecasting of rapid intensification) were needed, both NOAA and the nation began studies to determine how best to approach this. The National Science Board performed a detailed study in which both our new director of OAR, Dr. Rick Spinrad, and I gave presentations and participated in the discussions. Meanwhile, NOAA's Science Advisory Board formed a Hurricane Intensity Research Working Group, which also conducted a detailed study and made strong recommendations.

In response to the recommendations from both of these studies, NOAA held a summit in 2007 that included the directors of the National Weather Service and OAR, the director of the National Centers for Environmental Prediction, the director of the National Hurricane Center, and me. The discussion focused mostly on the improvement of hurricane intensity forecasting. I agreed strongly with this as a goal, but also argued for the inclusion of research to improve track and especially landfall predictions so that overwarning could be significantly reduced. This summit resulted in NOAA forming the Hurricane Forecast Improvement Project (HFIP).[20] Dr. Spinrad asked me if I would be willing to commit AOML's hurricane research to this project and I readily agreed.

HFIP was set up to provide a basis for NOAA, other agencies, and the larger scientific community to coordinate hurricane research. The overarching goals of HFIP are to improve the accuracy and reliability of hurricane forecasts, to extend the lead time for hurricane forecasts with increased certainty, and to increase

20. Gall et al., 2013.

confidence in hurricane forecasts so that fatalities and economic losses would be reduced substantially. At the outset, it was recognized that these efforts would require major investments in enhanced observational capabilities and strategies, and substantial improvements to data assimilation, numerical models, and forecast tools based on high-resolution and ensemble-based numerical prediction systems. The specific goals of HFIP (which were established in 2007) were to reduce the average errors of hurricane track and intensity forecast guidance by 20% within five years and 50% in 10 years, and also to extend the forecast period from five to seven days. Additional goals related to improving the prediction of rapid intensification (RI) of tropical cyclones, specifically to be able to more accurately predict when RI would occur, and also to reduce the overprediction of RI occurrence. We recognized at the outset that these were "stretch goals" and that they probably would be very difficult to achieve. Nevertheless, accelerated progress toward them would be of great societal benefit. One other goal that was added after the summit was to improve storm surge prediction. This came about after the National Weather Service made a decision to separate surge from the Saffir-Simpson hurricane wind scale and realized that they would need a specific storm surge product to replace the implied linkage to the wind scale. HFIP was asked to contribute support to the National Ocean Service–led storm surge effort.

After I returned to AOML from the summit, I briefed Frank Marks on what had transpired and he enthusiastically embraced the goals of HFIP. Even though HFIP did not officially begin and funding did not increase until 2009, we began preparatory work for this very important project. Frank became the scientific lead for HFIP and Dr. Ed Rappaport, the deputy director of the National Hurricane Center and an exceptional meteorologist, became the operational lead. Frank and Ed interacted regularly, while I interacted with Bill Proenza, the NHC director, all to establish improved collaboration between HRD and NHC. At the same time, we began the process of interacting more with NOAA's Environmental Modeling Center (EMC).

When I had first arrived at AOML, the Hurricane Research Division was performing primarily observational research and had developed tremendous expertise as well as extraordinary data sets for both research and operations.[21] However, with only a few exceptions most of the knowledge generated within HRD wasn't getting into the numerical models. The opportunity to change this for hurricanes came when Dr. D. B. Rao at the National Centers for Environmental Prediction (NCEP) contacted me regarding a very bright young numerical modeler that he

21. Aberson et al., 2006; Rogers, 2021.

knew was very interested in working on hurricane modeling at AOML. On my next trip to Maryland, I met with Dr. Sundararaman Gopalakrishnan (known to his friends and colleagues as Gopal). Gopal had been a key member of the team at NCEP that was developing NOAA's next hurricane forecasting model (known as HWRF) and had moved on to another organization where he would not be able to continue his hurricane research. I was immediately impressed with Gopal's abilities, motivation, and character and not long thereafter Gopal joined the staff of HRD. I knew from Gopal's past experience with HWRF that AOML would be able to make a significant contribution to the development of this model. I also saw this as a new opportunity to redefine the ways in which research is transitioned to operations. In the old way of doing this, research and operations worked separately, and there was what many consider a "valley of death" between them. As a result, much of the research that occurred never made it into operations. The approach I was planning not just for HRD, but for all of AOML's divisions was to get rid of the "valley of death" at the start by forming teams with both research and operations working together. This took a little time to get going, but once we formed the teams, the transition of research into operations occurred much more rapidly than would otherwise have been possible.

Initially, Frank and I began this process by forming a modeling group within HRD with Gopal as the lead scientist. Gopal, together with Dr. Xuejin Zhang and the rest of the modeling group, quickly demonstrated their ability to make significant contributions to HWRF. This led to forming a team research activity with EMC, and this, in turn, led to very substantial improvements to HWRF in terms of resolution, model physics, and infrastructure.[22]

AOML/HRD also contributed to achieving the HFIP goals through its unique observational research. This involved HRD scientists flying aboard NOAA Hurricane Hunter aircraft into numerous tropical cyclones at every stage of storm development. These flights provided extremely valuable information for the NHC forecasters, as well as for the EMC and HRD modelers to improve the HWRF model physics. The flights also provided a means to test and experiment with new observing systems. One that was provided to us by the Office of Naval Research was a Doppler wind lidar (DWL). Lidars are laser radars. Research that my PhD student Lisa Bucci conducted as part of her doctoral thesis demonstrated that DWL can effectively complement the tail Doppler radar aboard the Hurricane Hunter aircraft.[23]

22. Atlas et al., 2015c.
23. Bucci et al., 2018; Zhang et al., 2018.

In addition to the tremendous contributions to HFIP by HRD, AOML's Physical Oceanography Division (PhOD), led at this stage by Dr. Gustavo Goni, also made significant contributions. This included deploying autonomous ocean gliders in advance of storms, conducting numerical experiments, and having ocean scientists interact with hurricane scientists.[24]

My Involvement in AOML's Oceanographic Research

Although my interactions with AOML's open ocean and coastal ocean research were somewhat less than with HRD, I did interact very regularly with the PhOD and OCED directors and many of their staff on a variety of topics. A major activity that I initiated at AOML and with our University of Miami Rosenstiel School of Marine and Atmospheric Sciences (RSMAS) colleagues was to develop a capability to perform rigorous ocean OSSEs. Professor Chris Moors at RSMAS invited me to present a seminar on OSSEs shortly after I arrived at AOML. Chris and Silvia Garzoli then worked with me to organize an Ocean OSSE workshop involving multiple government agencies and academia. At the workshop, it was clear that previous ocean OSSEs that had been conducted by the oceanographic community had not followed the steps necessary to ensure realism and credibility. Everyone at the workshop agreed that there was a strong need to improve how ocean OSSEs are conducted, but there was also skepticism about whether a credible ocean OSSE system could be developed given the current state of ocean modeling. Following this workshop, we hired Dr. George Halliwell at AOML to work on ocean modeling and to begin to develop an ocean OSSE system first for the Gulf of Mexico, later to be expanded to the entire North Atlantic, and ultimately to cover all of the global oceans. This activity was very successful, and AOML developed the first in the world rigorous ocean OSSE system following the procedures that we had previously developed for numerical weather prediction OSSEs. Numerous OSSEs have already been conducted with this system, and as a result of the journal articles[25] and presentations that George and Professor Villy Kourafalou of the University of Miami have given, this improved ocean OSSE methodology is now being used worldwide. In addition, work at AOML in OCED was initiated to develop an OSSE system for marine ecosystems and fisheries applications.

Another area in which I have interacted strongly with the ocean research at AOML is with ocean surface winds. In particular, during my first few years at

---

24. Dong et al., 2017.
25. Halliwell et al., 2014, 2015, 2017a,b; Oke et al., 2015; Androulidakis et al., 2016.

AOML I continued to lead the CCMP ocean surface wind data set that I began at NASA. Several oceanographers in PhOD and OCED made use of this data set in their research. They included the outstanding ocean carbon researcher Dr. Rik Wanninkhof and his colleagues. I also became involved with several of the climate research activities in PhOD and often stimulated new research studies in the same way Milt Halem and Eugenia Kalnay had done for me many years earlier. One of these studies dealt with the ocean's influence on major tornado outbreaks over the United States. This study, led by Dr. Sang-Ki Lee, resulted in the development of an experimental seasonal tornado outlook. Other studies dealt with the ocean's role in heat waves and hurricanes.

One additional area of research that I supported strongly was the environmental microbiology work being conducted in OCED, even though this was an area that I had never studied before. From my first interactions with Drs. Kelly Goodwin, Chris Singiliano, Maribeth Gidley, and also John Proni, I became convinced that this was extremely important research for the nation and for local communities. During my tenure as AOML director, there was a significant expansion of this environmental microbiology program, which utilized a wide range of traditional bacteriological methods while also developing, testing, and deploying cutting-edge molecular technologies. These approaches led to a better understanding of marine microbiomes and their ecological functions. Other benefits included the ability to detect and track sources of microbial contaminants from land-based sources of pollution, to enhance water quality assessments, and to help guide management to improve water quality. The research projects covered a wide variety of topics including coral health, marine animal health, water quality, microbial contaminants and pathogens at beaches and recreational waters, marine vectors of disease transmission, and observations of marine biodiversity.

The Formation of NOAA's Quantitative Observing System Assessment Program

Not long after I arrived at AOML, I realized that many observing system decisions in NOAA were being made based on subject matter experts' opinions rather than on the basis of numerical experiments. To begin to change this, I created an observing system simulation experiment (OSSE) testbed. The primary objectives were to establish a numerical testbed that would enable a hierarchy of experiments to (1) determine the potential impact of proposed space-based, suborbital, and surface-based observing systems on analyses and forecasts, (2) evaluate trade-offs in observing system design, (3) assess proposed methodologies for assimilating new observations, in coordination with the Joint Center for Satellite

Data Assimilation (JCSDA), and (4) provide a place to evaluate the impacts of changes in observing system capabilities due to either system degradation or proposed capabilities. Subobjectives were to define both the advantages and limits of different types of numerical experiments and tools for this purpose and to generate an OSSE process that invites participation by the broad community of agency planners, research scientists, and operational centers. The OSSE testbed helped to support the initial development of a hurricane OSSE system that has been used for a wide variety of very useful predictability and observing system experiments for hurricanes and severe local storms.

In 2013, Dr. Alexander (Sandy) MacDonald and I proposed the formation of a broader effort, which we called the Quantitative Observing System Assessment Program (QOSAP). Initially, there was strong resistance to this, but a year later QOSAP began and absorbed the OSSE testbed. The primary objective of QOSAP was to increase the use of quantitative assessments for proposed changes to the composite observing system. In a time of tight budgets, quantitative observing assessments enable better predictions at lower cost. Our plan was to create the capability to perform these assessments for all global and national observing systems for physical, chemical, and biological aspects of the ocean and atmosphere. It was thus set up to have participation from all of NOAA's line offices.

Following the establishment of QOSAP, numerous OSSEs were performed for nearly all of NOAA's line offices, as well as for other agencies. These included global and regional weather prediction OSSEs to evaluate a variety of satellite and aircraft observing systems, regional OSSEs to investigate different aspects of hurricane predictability, and ocean OSSEs to evaluate the deployment and design of ocean observing systems for both meteorological and oceanographic applications. Specific OSSEs were conducted to support several new satellite missions, determine the relative value of different technologies for wind profiling from space, investigate various scenarios for the deployment of NOAA's Hurricane Hunter aircraft, assess the potential for an enhanced constellation of radio occultation satellites, and determine observational requirements for reducing systematic errors in regional hurricane models.[26] As the director of QOSAP, I also served on NOAA's Space Platform Requirements Working Group (SPRWG), which was led by Dr. Rick Anthes. The SPRWG worked to establish priorities for NOAA's satellites in the 2030–2050 time frame.[27]

26. Atlas et al., 2015a,b; Annane et al., 2017; Leidner et al., 2018; Cucurull et al., 2018; Ryan et al., 2019.

27. Anthes et al., 2019.

In 2016, Dr. Ross Hoffman and I published an article on the future of OSSEs in the *Bulletin of the American Meteorological Society*.[28] As a supplement to this article, we included an interactive checklist that detailed the rules and procedures for how OSSEs should be conducted to ensure realism and credibility so that these types of experiments could be designed and conducted correctly by the larger scientific community.

QOSAP quickly became an active and very appreciated activity. In 2017 Congress passed the Weather Forecast Improvement Act. This act was the first time that legislation recognized the importance of OSSEs and other quantitative measures in optimizing the observing system for improved forecasting at lower cost.[29] In 2019 I turned over the scientific leadership of QOSAP to Dr. Lidia Cucurull, who had been very effectively serving as QOSAP's deputy director and chief scientist.

Additional Thoughts on Management and Leadership

Over the nearly 14 years that I served as director of AOML, I encountered a wide range of management and leadership issues. At the outset, I'd like to say that being a manager and being a leader are very different. Leaders provide direction to an organization and inspire their employees. True leaders also care more about their employees' careers than they do about their own. To be an effective leader of a scientific research laboratory or center, it is very important to have been an active scientist first, and I believe that it is also important to continue doing at least some research while in the leadership position. By continuing to do research, the leader is able to identify more closely with the issues and concerns faced by the scientific staff of the organization. The leader is thus able to help guide the staff more productively so that they achieve their full potential, and the mission of the organization is more effectively achieved. This is how I attempted to function as AOML's director. I continued to do research on a variety of topics and continued to give presentations at conferences and publish in scientific journals. However, I was always happiest and most effective as director when younger employees would take ownership of a research activity and be the first author on publications. When I joined OAR in 2005, there were other laboratory directors who functioned in a similar way, and I remember feeling very glad to have joined such an elite group of scientific leaders.

---

28. Hoffman and Atlas, 2016a,b.
29. Zheng et al., 2020.

Other policies that are essential to having a strong and productive scientific organization are having an "open" organization that welcomes and encourages external collaboration, having an "open door" policy so employees can easily bring issues that concern them to their leadership, showing appreciation for employees and colleagues, having strong interactions with stakeholders, and having the right people in key positions.

By having an open organization during my tenure as AOML director, collaborations increased substantially and this, in turn, increased AOML's productivity in terms of both publications and transitions of research. AOML's publications in refereed journals doubled, and our research transitions increased severalfold during this time period. Having an open-door policy benefited both the employees and me. Although I sometimes had to deal with complaints and problems, I also really enjoyed the interactions with employees, especially when they wanted to discuss their scientific results. I found that I also really enjoyed showing appreciation for all of the employees at AOML, but such appreciation has to be genuine to be effective. Finally, interacting with stakeholders is an essential task for the leader of any organization. For AOML, our stakeholders were the other NOAA line offices, the U.S. Congress, and in an indirect way the general public. By forming teams and sharing and transitioning research, we served NOAA's operational line offices. For Congress, all of the OAR lab directors briefed our local representatives and senators regularly. I found that speaking honestly with them about what we were doing at AOML and how it benefits their constituents and the nation, answering their requests in a timely manner, and also participating with them in hurricane preparedness briefings was always greatly appreciated. For the general public, outreach and education are essential. At AOML, Erica Rule led this activity during my tenure as director, and together with Evan Forde and several others, they created strong ties with the South Florida community. Several people at AOML and I would sometimes teach at local schools, and AOML would also host an open house that was always very well attended and appreciated. I also found that besides being interested in what we were doing to improve forecasting and protect the environment, the public was also very appreciative of how careful we were with our budgets and how we always strived to keep costs as low as possible.

One of the most interesting and demanding challenges that I faced occurred within my first few years at AOML. At this time, there was what the media referred to as a mutiny at the National Hurricane Center. I had become close friends with the director of NHC, as we had worked together to initiate HFIP, and I knew him to be a dedicated public servant and a sincere individual. The challenge was how to stand up for him without weakening the relationship with the forecasters at NHC. I received very good advice from my leadership, Drs. Spinrad and MacDonald, just before I was to testify to Congress on his behalf. They advised me to stick to the science, which I did. The outcome from the hearing was very positive and AOML's relationship with NHC continued to improve at an even faster pace.

Many other interesting challenges occurred while I was the director of AOML. These included several government shutdowns, more hurricanes hitting Miami, and major oil spills. I'm proud to say that in every case AOML employees rose to the occasion and performed in an exceptional manner. This was due in large part to having the right people in key positions. During my last several years at AOML, Dr. Molly Baringer, an outstanding ocean scientist, served as the lab's deputy director. Dr. Gustavo Goni, who consistently set an example of high productivity, replaced Silvia Garzoli as the director of PhOD. He and his deputy, Dr. Rick Lumpkin, continued the tremendous legacy of Dr. Garzoli and led PhOD to significant advances in ocean and climate research. Dr. Jim Hendee as director of OCED, his deputy Dr. Chris Kelble, and senior scientist Dr. Rik Wanninkhof led their division to significant advances in understanding ecosystems and the effects of climate change on coral reefs. Finally, Dr. Frank Marks and his deputy Shirley Murillo continued the outstanding legacy of HRD in advancing hurricane science.

Chapter 6

Teaching at Colleges and Universities

I have been very fortunate throughout my entire research career in that I was able to teach almost continuously at nearby colleges and universities. This was mostly on a part-time basis as an adjunct instructor, and in later years as an adjunct professor. Following my first college teaching experience at the State University of New York (SUNY) Agricultural and Technical College in 1973, I was next hired to teach at the SUNY Maritime College, and I taught there from 1974 to 1977. As the name suggests, the Maritime College was a very specialized school. There I taught their synoptic meteorology and weather forecasting courses. The Maritime College had well-prepared curricula and exercises, but I was also able to add to the course from my experience as a forecaster.

I next taught first part-time and then full-time for one year at SUNY Stony Brook, and also part-time at Adelphi University. At Stony Brook, a few others and I were teaching atmospheric science courses in the Mechanical Engineering Department. For reasons I no longer remember, nearly all of the courses were taught without prerequisites. The freshman course on weather and climate, which was taught by a different professor, had over 800 students enrolled. The sophomore course on elementary geology, meteorology, and oceanography that I taught had over 200 students. The senior-level weather prediction course had 60 students the first time I taught it. The next time I taught the weather prediction course, there were over 90 students, which made lab work very challenging. Even the graduate course that I taught on advanced synoptic meteorology had many more students than is typical. We also created the Stony Brook Weather Observatory and I served as its director. At the Weather Observatory, students made forecasts and participated in research projects. During the winter, forecasts of the probability of snow amounts in different categories were issued. These were used and very much appreciated by the local highway department.

Many of the best students at Stony Brook were hired to work at NASA GISS while I was there, and several of them continued to work at NASA GSFC when the Global Weather Group at GISS moved there. Others went to work for the National Weather Service, private firms, or in academia. However, not all of the students were interested in becoming meteorologists. Many just had a strong interest in the weather and wanted to learn as much about it as they could while attending college. Some of these were students in other scientific disciplines, but others had no scientific or mathematical background. Teaching this diverse group was sometimes very challenging. For example, a drama student who had never had calculus, let alone differential equations, failed the midterm exam in one of my courses. He came to me for help, and I worked with him so that he could gain a sufficient understanding of the mathematics. I remember being very pleased when he successfully derived equations and earned a solid B on the final exam for the course.

After moving to Maryland to become a civil servant at NASA's Goddard Space Flight Center in 1978, I was asked to teach at the University of Maryland at College Park (UMCP) in their Department of Meteorology and Oceanography. I did so for nearly all of the 27 years that I was at Goddard. On occasion, I also taught at the University of Maryland in Baltimore County (UMBC) and at Catonsville Community College (CCC). At UMBC I taught a course on weather and climate in their Geography Department, while at CCC I taught a course on weather prediction primarily for pilots and others in the aviation field.

The first course I taught at UMCP was a two-semester undergraduate general meteorology course that covered core topics in synoptic, dynamic, and physical meteorology. It was nice to once again have all of the students with the proper prerequisites in math, physical science, and computer science, and I found the UMCP students to be very well prepared. I started the course with the history of atmospheric science, and then covered the basics of weather and weather systems, followed by surface and upper-air weather map analysis. I then reviewed all of the essential mathematical concepts and covered the fundamental equations of atmospheric dynamics. For each of the mathematical operators and for each of the equations, I showed their relationship to what we see on weather maps. The goal here was to enable students to think interchangeably between the maps and equations. Physical meteorology was covered next. This included the nature of radiation and heat transfer, cloud and precipitation processes, and thermodynamics. Finally, we made a more in-depth study of weather systems and weather forecasting.

After the first year, I regularly taught the weather prediction course at UMCP. In this course, I covered the history, evolution, and essential nature of weather forecasting, followed by specific forecasting methods for cyclogenesis (the formation of new cyclones), the movement and intensity changes of cyclones, anticyclones, surface fronts, and upper-level features, the forecasting of each of the weather elements, and the application of numerical models. This latter aspect covered not only how to use and interpret the numerical model output, but also how to modify the model predictions or determine if the solution from one particular model is more credible than that of other models. This was done by utilizing observational data that was not assimilated directly into a model (such as satellite imagery), evaluating how well the initial conditions for the model represented the observations that were assimilated (e.g., looking for small-scale features that might have been smoothed out by the assimilation process or not adequately resolved by the model), and correcting for other model deficiencies and systematic and nonsystematic errors. This approach worked very well through the early 2000s, but is somewhat more difficult to apply with current very high-resolution models. With today's models it is still very important for forecasters to understand the strengths and weaknesses of each of the models that they use.

In my weather prediction courses, I also taught empirical methods as a way of rapidly building forecast experience. However, with each of the methods that I taught, I would also explain the dynamical processes associated with the empirical technique and how they could be utilized in conjunction with the numerical forecast models. A good example of this is the cyclogenesis prediction method described by J. J. George in his 1960 book *Weather Forecasting for Aeronautics*.[1] In this method, a forecaster looks for the occurrence of a "cold air injection" (CAI), which is another name for a low-level jet that transports cold air in a narrow current. In the empirical method, if a CAI is observed, the forecaster goes through a detailed procedure to forecast the occurrence of either cyclogenesis, a center jump, or intensification of an already existing cyclone. The forecaster would then go on to determine the location, intensity, and speed and direction of movement for the cyclone.

1. George, 1960.

Two of my students and I conducted a study to determine the climatological characteristics of cold air injections[2] and also to investigate how forecasters could use aspects of the empirical method in conjunction with numerical model predictions. The study found that one of the three alternatives (cyclogenesis, a center jump, or intensification of an existing low-pressure system) follows a CAI only 70% of the time. To determine what will actually occur, it is necessary to understand the dynamics associated with the CAI. CAIs can deform the thermal structure significantly and are typically associated with rising sea level pressures below and decreasing upper-level heights above it. To use the occurrence of the CAI effectively it is thus necessary to determine its relationship to both surface and upper-level meteorological features. For example, if the CAI is located directly under a somewhat broad upper-level trough, then sharpening of the trough is likely to result, and this can enhance the upper-level forcing for surface pressure falls and cyclonic development in advance of the trough and/or cause a change in how a surface feature would be steered by the upper-level flow. This was a fairly common cause of nonsystematic errors in numerical model predictions until such time as CAIs could be accurately represented.

Teaching synoptic meteorology at the graduate and undergraduate levels was a source of great pleasure for me throughout my research career. During this time, I taught more than a thousand students. Many became meteorologists with the National Weather Service and rose to high positions. Others became research scientists, television meteorologists, college and university professors, and some started their own companies as part of the Weather Enterprise. My last year teaching was 2003. A few weeks into the semester, I broke both of my arms in a bicycle accident. I wasn't sure what to do until my wife volunteered to go with me to each remaining class and to be my arms. My wife, who had been a high school math teacher, was soon asking questions in class along with the registered graduate students. This was especially enjoyable for me and made my last teaching experience at UMCP very special.

Like most synoptic meteorology professors, I often used weather situations on which I had conducted research to illustrate important points, concepts, or methods. I encouraged the students to go through a reasoning process similar to the evolution of forecasting itself. This meant starting from local observations, then making a detailed study of the surface weather map, then utilizing the surface and upper-air maps in combination, and only then making a detailed evaluation of what the available numerical models were predicting.

2. Atlas et al., 1980.

The final step was to reconcile their reasoning and calculations with the numerical models to produce a forecast.

The February 1976 case that I had studied during the NASA/NOAA Data Systems Tests was one example of such a case. Besides illustrating the dynamics of cyclone movement and intensification and the development of different types of severe weather, this case also illustrated one of the well-known ways in which satellite imagery could be used by a forecaster in conjunction with numerical models. In this case, there were two numerical forecasts for the students to consider—one that had assimilated satellite temperature soundings and one that did not. The initial conditions for these forecasts were very different in their representation of the vorticity and vorticity advection (transport of vorticity by the wind) off the west coast of the United States. The satellite imagery at this time showed a comma-shaped cloud that is consistent with the forecast that included satellite temperature sounding data. This and the synoptic and dynamic reasoning that the student forecaster would have gone through in this case enables the forecaster to decide between the two model solutions.

Another of the many forecasting examples that I used in my teaching was the President's Day Cyclone of 1979. As mentioned earlier, this was a very significant storm that produced greater than two feet of snow in the middle Atlantic states and crippled the Washington, DC, area. As documented by Drs. Bosart, Uccellini, and me, the operational predictions for this storm were poor and greatly underestimated the amount of snow that would fall. This was in part due to the poor predictions of the limited area fine mesh (LFM) numerical model, which was operational at that time. The 36-hour LFM forecast from 0000 GMT 18 February 1979 failed to predict the occurrence of a cyclone near the mid-Atlantic coast, while the 12- and 24-hour predictions forecast the cyclone to be too weak and too far south of its observed position. Nevertheless, there were strong indicators for the development of an intense cyclone off the east coast of the United States and for it to follow a track similar to what was observed. I used this case in my weather prediction classes to illustrate one of the ways a forecaster might have good indications of the development and movement of this storm, even though the available numerical model predictions were less than adequate. A portion of this reasoning will be shown here.[3]

Using the surface, 850 mb, and 500 mb maps that were available before and during the storm, I would ask the students first if there were any factors

3. The weather maps for this case are available in Atlas and Rosenberg (1982) and Kocin and Uccellini (2004).

at 0000 GMT on February 18, 1979 that are indicative of the potential for the development of a strong coastal cyclone on February 19. These would be the extremely cold Arctic air mass that was over the northeastern United States at this time and the unusually strong advection of cold dry air over the warm waters of the Gulf Stream. This would result in very strong heating of the air above. Another factor would be a short wave trough at 500 mb that while still far away from the east coast of the United States, was moving toward this area. A third factor would be a weak cold air injection at 850 mb, and a fourth would be the surface front over the southern United States and Gulf of Mexico. I then would ask the students what factors at 1200 GMT 18 February would contribute to the development of a shallow low-pressure system east of South Carolina on 0000 GMT 19 February and why this cyclone would not be stronger at this time. One important factor for the shallow cyclone development would be very strong low-level warm air advection (transport of lighter warm air by the wind) over this area, which was evident on the 850 mb chart. The cyclone would not be much stronger at this time because the 500 mb trough was not yet approaching this area and there was little upper-level support for development. Using the next sets of maps, I then would ask the students why the cyclone appears to move northward rather than further out to sea from 0000 to 1200 GMT on February 19. This is because there was both strong warm advection (evident at 850 mb) and strong upper-level forcing (evident at 500 mb) to the north of the cyclone. Finally, I would ask the students why the surface cyclone moves eastward and continues to intensify after 1200 GMT on February 19. This was due to the eastward movement of the upper-level trough and the strong warm advection to the east of the surface low at this time.

Although the discussion above has been simplified for this memoir, it should be clear that a forecaster should continue to use meteorological reasoning in conjunction with numerical models. At times, such reasoning can enable the forecaster to make more effective use of numerical model guidance, either by gaining confidence in a model's predictions, modifying the model's forecasts, or by choosing to believe one model's forecasts over another. For the interested reader, many more important details of this storm are in the journal articles published by Drs. Bosart, Uccellini, and me and in the excellent book on east coast snowstorms by Kocin and Uccellini.

Chapter 7

Retirement

The decision to retire was a very difficult one for me. I loved being the director of AOML, and the people there had become like family to me. However, I felt that retirement would be a challenge for me, and I've always liked taking on challenges. I also felt that I had such an exceptional deputy at AOML that she deserved a chance to become the director. So with the idea of spending more time with my wife, son, daughter, and grandchildren I set a date of February 2, 2019 to retire. As it turned out, the government was going into a shutdown as we neared the end of 2018. I felt it would be wrong to abandon my post during such an event, so I called one more all-hands meeting at AOML to brief the staff on my decision to postpone my retirement by one month, and also to prepare them for the impending shutdown.

With the exception of the shutdown, my last few months at AOML were very special. At an OAR Senior Research Council meeting, there was a modest retirement celebration for me and another lab director. Then on February 19, I was honored on the floor of the U.S. House of Representatives. This was followed by a celebration at the National Hurricane Center on February 21 where they presented me with an award for enduring contributions to the nation's hurricane forecast and warning program. On February 26, the University of Miami held a science symposium in my honor and a retirement celebration. Here, many of my friends, colleagues, and former students participated. These included Milt Halem, Eugenia Kalnay, Jagadish Shukla, Ross Hoffman, Shian-Jiann Lin, Arlindo DaSilva, Bill Putman, Bob and Jean Rosenberg, and Ben Kirtman, the director of the Cooperative Institute for Marine and Atmospheric

Studies and the host for the event. In addition, Director Louis Uccellini of the National Weather Service, Deputy Director Rear Admiral Nancy Hahn of the NOAA Corps, Director Dr. Venkatachalam Ramaswamy of NOAA's Geophysical Fluid Dynamics Laboratory, Director Dr. Robert Webb of NOAA's Earth System Research Laboratory, the dean and faculty from the University of Miami RSMAS, and nearly all of the staff from AOML also attended. From my family, my son Ken and my wife Shirley attended, making the event very special. Finally, on March 1, 2019, my last day of work as a federal scientist, AOML hosted a farewell party for me. At this last event, I was presented with an official proclamation that made me the first director emeritus of AOML.

With this, I entered the world of retirement, but my retirement was from the federal government and not from being a scientist. Since March 2, 2019, I've still been very busy and continue to do many activities that are a rewarding part of a scientist's career. Along with Professor Sharanya Majumdar, I served as co-advisor to an excellent young scientist as she completed her PhD thesis. Also with Professor Majumdar, I continue to serve as chair of the American Meteorological Society's Conference on Integrated Observing and Assimilation Systems for the Atmosphere, Ocean, and Land Surface (IOAS-AOLS). Since retiring, I've co-authored 12 peer-reviewed articles in scientific journals, reviewed innumerable papers for journals, served on several scientific panels, provided scientific advice to many colleagues in NASA and NOAA, provided forecasts and advice on hurricanes to family and friends, and interacted with several university colleagues on important research topics. I've also guest lectured in person at the University of Miami and remotely for the University of Maryland, presented the keynote address at the International Winter Simulation Conference, and have begun new interactions with faculty at UMBC and other universities. Finally, I am also continuing to learn, because even after 60 years in meteorology there is still much to learn.

References

Aberson, S.D., M.L. Black, R.A. Black, R.W. Burpee, J.J. Cione, C.W. Landsea, and F.D. Marks, 2006: Thirty years of tropical cyclone research with the NOAA P-3 aircraft. *Bulletin of the American Meteorological Society*, 87 (8), 1039–1055.

Androulidakis, Y.S., V.H. Kourafalou, G.R. Halliwell, M. Le Henaff, H.S. Kang, M. Mehari, and R. Atlas, 2016: Hurricane interaction with the upper ocean in the Amazon-Orinoco plume region. *Ocean Dynamics,* 66 (12), 1559–1588.

Annane, B., B. McNoldy, S.M. Leidner, R. Hoffman, R. Atlas, and S.J. Majumdar, 2018: A study of the HWRF analysis and forecast impact of realistically simulated CYGNSS observations assimilated at scalar wind speeds and as VAM wind vectors. *Monthly Weather Review,* 146 (7), 2221–2236.

Anthes, R., S. Ackerman, R. Atlas, L.W. Callahan, G.J. Dittberner, R. Edwing, P. Emch, M. Ford, W.B. Gail, M. Goldeberg, S. Goodman, C. Kummerow, M.W. Mair, T. Onsager, K. Schrab, T. von der Haar, and J. Yoe, 2019: Developing priority observational requirements from space using multi-attribute utility theory. *Bulletin of the American Meteorological Society,* 100 (9), 1753–1793.

Atlas, R., 1975: Numerical experiments to investigate the role of horizontal temperature advection in a predictive mixed layer ocean model. *Goddard Institute for Space Studies Meteorology Research Review,* 78–85.

Atlas, R., 1978: The development of a computerized procedure for the prediction of severe local storm potential. *Atmospheric and Oceanographic Research Review, NASA Tech. Mem. 80253,* 20–24.

Atlas, R., 1979: A comparison of GLAS SAT and NMC high resolution NOSAT forecasts from 19 and 11 February 1976. *NASA Tech. Mem. 80591.*

Atlas, R., M. Halem, and M. Ghil, 1979: Subjective evaluation of the combined influence of satellite temperature sounding data and increased model resolution on numerical weather forecasting. *Proceedings, Fourth Conference on Numerical Weather Prediction,* Oct. 29–Nov. 1, 1979, Silver Spring, MD, 319–328.

Atlas, R., R. Rosenberg, and G. Cole, 1980: A comparison between an empirical technique for the prediction of cyclogenesis and the LFM II. *Proceedings, Eighth Conference on Weather Forecasting and Analysis*, June 10–13, Denver, CO, 435–441.

Atlas, R., M. Ghil, and M. Halem, 1981: Reply to comment by L. Druyan on time-continuous assimilation of remote-sounding data and its effect on weather forecasting. *Monthly Weather Review*, 109 (1), 201–204.

Atlas, R., 1982: The growth of prognostic differences between GLAS model forecasts from SAT and NOSAT initial condition. *Monthly Weather Review*, 110 (7), 877–882.

Atlas, R., M. Halem, and M. Ghil, 1982: The effect of model resolution and satellite sounding data on GLAS model forecasts. *Monthly Weather Review*, 110 (7), 662–682.

Atlas, R., and R. Rosenberg, 1982: Numerical prediction of the mid-Atlantic states cyclone of 18–19 February 1979. *NASA Tech. Mem. 83992*, 53 pp.

Atlas, R., E. Kalnay, J. Susskind, W.E. Baker, and M. Halem, 1984: Simulation studies of the impact of advanced observing systems on numerical weather prediction. *Proceedings, Conference on Satellite Meteorology/Remote Sensing Applications*. June 25–29, Clearwater Beach, FL, 283–287.

Atlas, R., E. Kalnay, W.E. Baker, J. Susskind, D. Reuter, and M. Halem, 1985a: Simulation studies of the impact of future observing systems on weather prediction. *Proceedings, 7th AMS Conference on Numerical Weather Prediction*, June 17–20, Montreal, Canada, 145–151.

Atlas, R., E. Kalnay, and M. Halem, 1985b: Impact of satellite temperature sounding and wind data on numerical weather prediction. *Optical Engineering*, 24 (2), 341–346.

Atlas, R., 1987: The role of oceanic fluxes and initial data in the numerical prediction of an intense coastal storm. *Dynamics of Atmospheres and Oceans*, 10 (4), 359–388.

Atlas, R., A.J. Busalacchi, E. Kalnay, S. Bloom, and M. Ghil, 1987: Global surface wind and flux fields from model assimilation of Seasat data. *Journal of Geophysical Research*, 92 (C6), 6477–6487.

Atlas, R., N. Wolfson, and Y.C. Sud, 1988: Numerical prediction experiments related to the summer 1980 U.S. heat wave. *Proceedings, Eighth Conference on Numerical Weather Prediction*, 719–725.

Atlas, R., and G.D. Emmitt, 1991: Implications of several orbit inclinations for the impact of LAWS on global climate studies. *Proceedings, AMS Second Symposium on Global Change Studies*, January 14–18, New Orleans, LA, 28–32.

Atlas, R., N. Wolfson, and J. Terry, 1993: The effect of SST and soil moisture anomalies on GLA model simulations of the 1988 U.S. summer drought. *Journal of Climate*, 6 (11), 2034–2048.

Atlas, R., R.N. Hoffman, S.C. Bloom, J.C. Jusem, and J. Ardizzone, 1996: A multiyear global surface wind velocity dataset using SSM/I wind observations. *Bulletin of the American Meteorological Society*, 77 (5), 869–882.

Atlas, R., 1996a: Application of SSM/I wind speed data to weather analysis and forecasting. *Proceedings, 15th Conference on Weather Analysis and Forecasting*, 138–141.

Atlas, R., 1996b: The impact of ERS-1 scatterometer data on GEOS and NCEP model forecasts. *Proceedings, 11th Conference on Numerical Weather Prediction,* 99–101.

Atlas, R., 1997: Atmospheric observations and experiments to assess their usefulness in data assimilation. *Journal of the Meteorological Society of Japan,* 75 (1B), 111–130.

Atlas, R., S.C. Bloom, R.N. Hoffman, E. Brin, J. Ardizzone, J. Terry, D. Bungato, and J.C. Jusem, 1999: Geophysical validation of NSCAT winds using atmospheric data and analyses. *Journal of Geophysical Research,* 104 (C5), 11,405–11,424.

Atlas, R., R.N. Hoffman, S.M. Leidner, J. Sienkiewicz, T.-W. Yu, S.C. Bloom, E. Brin, J. Ardizzone, J. Terry, D. Bungato, and J.C. Jusem, 2001: The effects of marine winds from scatterometer data on weather analysis and forecasting. *Bulletin of the American Meteorological Society,* 82 (9), 1965–1990.

Atlas, R., R.O. Reale, B.-W. Shen, S.-J. Lin, J.-D. Chern, W. Putman, T. Lee, K.-S. Yeh, M. Bosilovich, and J. Radakovich, 2005: Hurricane forecasting with the high-resolution NASA finite volume general circulation model. *Geophysical Research Letters,* 32 (3), L03807, https://doi.org/10.1029/2004GL021513.

Atlas, R., S.-J. Lin, B.-W. Shen, O. Reale, and K.-S. Yeh, 2007: Improving hurricane prediction through innovative global modeling. In *Extending the Horizons: Advances in Computing, Optimization, and Decision Technologies,* E.K. Baker, A. Joseph, A. Mehrotra, and M.A. Trick (eds.). Springer, 1–14.

Atlas, R., and G.D. Emmitt, 2008: Review of observing system simulation experiments to evaluate the potential impact of lidar winds on numerical weather prediction. *ILRC24,* Vol. 2 (ISBN 978-0-615-21489-4), 726–729.

Atlas, R., and L.P. Riishojgaard, 2008: Application of OSSEs to observing system design. In *Remote Sensing System Engineering,* P.E. Ardanuy and J.J. Puschell (eds.). *Proceedings, SPIE,* 7087:708707, https://doi.org/10.1117/12.795344, 9 pp.

Atlas, R., R.N. Hoffman, J. Ardizzone, S.M. Leidner, J.C. Jusem, D.K. Smith, and D. Gombos, 2011: A cross-calibrated, multi-platform ocean surface wind velocity product for meteorological and oceanographic applications. *Bulletin of the American Meteorological Society,* 92 (2), 157–174.

Atlas, R., and T. Pagano, 2014: Observing system simulation experiments to assess the potential impact of proposed satellite instruments on hurricane prediction. *Proceedings of SPIE Imagery Spectrometry XIX,* August 18, 2014, San Diego, CA, 1–9.

Atlas, R., L. Bucci, B. Annane, R. Hoffman, and S. Murillo, 2015a: Observing system simulation experiments to assess the potential impact of new observing systems on hurricane forecasting. *Marine Technology Society Journal,* 49 (6), 140–148.

Atlas, R., R.N. Hoffman, Z. Ma, G.D. Emmitt, S.A. Wood, S. Greco, S. Tucker, L. Bucci, B. Annane, and S. Murillo, 2015b: Observing system simulation experiments (OSSEs) to evaluate the potential impact of an optical autocovariance wind lidar (OAWL) on numerical weather prediction. *Journal of Atmospheric and Oceanic Technology,* 32 (9), 1593–1613.

Atlas, R., V. Tallapragada, and S. Gopalakrishnan, 2015c: Advances in tropical cyclone intensity forecasts. *Marine Technology Society Journal,* 49 (6), 149–160.

Baker, W.E., R. Atlas, E. Kalnay, M. Halem, P.M. Woiceshyn, and D. Edelmann, 1984: Large-scale analysis and forecast experiments with wind data from the Seasat-A scatterometer. *Journal of Geophysical Research,* 89, 4927–4936.

Baker, W.E., G.D. Emmitt, F. Robertson, R.M. Atlas, J.E. Molinari, D.A. Bowdle, J. Paegle, R.M. Hardesty, R.T. Menzies, T.N. Krishnamurti, R.A. Brown, M.J. Post, J.R. Anderson, A.C. Lorenc, and J. McElroy, 1995: Lidar measured winds from space: A key component for weather and climate prediction. *Bulletin of the American Meteorological Society,* 76 (6), 869–888.

Baker, W.E., R. Atlas, C. Cardinali, A. Clement, G.D. Emmitt, B.M. Gentry, R.M. Hardesty, E. Kallen, M.J. Kavaya, R. Langland, M. Masutani, W. McCarty, R.B. Pierce, Z. Pu, L.P. Riishojgaard, J. Ryan, S. Tucker, M. Weissmann, and J.G. Yoe, 2014: Lidar-measured wind profiles: The missing link in the global observing system. *Bulletin of the American Meteorological Society,* 95 (4), 543–564.

Bloom, S.C., R. Atlas, and E. Brin, 1996: Validation of NSCAT data at GLA. *Proceedings, 15th Conference on Weather Analysis and Forecasting,* 135–137.

Bosart, L., 1981: The President's Day snowstorm of 18–19 February 1979: A subsynoptic-scale event. *Monthly Weather Review,* 109, 1542–1566.

Bucci, L.R., C. O'Handley, G.D. Emmitt, J.A. Zhang, K. Ryan, and R. Atlas, 2018: Validation of an airborne Doppler wind lidar in tropical cyclones. *Sensors,* 18 (12), 4288, 1–15.

Chahine, M.T., T.S. Pagano, H.H. Aumann, R. Atlas, C. Barnet, J. Blaisdell, L. Chen, M. Divakarla, E.J. Fetzer, M. Goldberg, C. Gautier, S. Granger, S. Hannon, F.W. Irion, R. Kakar, E. Kalnay, B.H. Lambrigtsen, S.-Y. Lee, J. LeMarshall, W.W. McMillan, L. McMillin, E.T. Olsen, H. Revercomb, P. Rosenkranz, W.L. Smith, D. Staelin, L.L. Strow, J. Susskind, D. Tobin, W. Wolf, and L. Zhou, 2006: AIRS: Improving weather forecasting and providing new data on greenhouse gases. *Bulletin of the American Meteorological Society,* 87 (7), 911–926.

Conaty, A.L., J.C. Jusem, L. Takacs, D. Keyser, and R. Atlas, 2001: The structure and evolution of extratropical cyclones, fronts, jet streams, and the tropopause in the GEOS general circulation model. *Bulletin of the American Meteorological Society,* 82 (9), 1853–1867.

Cucurull, L., R. Atlas, R. Li, M.J. Mueller, and R.N. Hoffman, 2018: An observing system simulation experiment with a constellation of radio occultation satellites. *Monthly Weather Review,* 146 (12), 4247–4259.

Dong, J., R. Domingues, G. Goni, G. Halliwell, H-S. Kim, S-K. Lee, M. Mehari, F. Bringas, J. Morell, and L. Pomales, 2017: Impact of assimilating underwater glider data on Hurricane Gonzalo (2014) forecasts. *Weather and Forecasting,* 32, 1144–1159.

Duffy, D., R. Atlas, T. Rosmond, E. Barker, and R. Rosenberg, 1984: The impact of Seasat scatterometer winds on the Navy's operational model. *Journal of Geophysical Research,* 89, 7238–7244.

Duffy, D., and R. Atlas, 1986: The impact of Seasat-A scatterometer data on the numerical prediction of the QE II storm. *Journal of Geophysical Research*, 91 (C2), 2241–2248.

Fisher, R.M., 1953: *How to Know and Predict the Weather*. New American Library, 167 pp.

Gall, R., J. Franklin, F. Marks, E.N. Rappaport, and F. Toepfer, 2013: The Hurricane Forecast Improvement Project. *Bulletin of the American Meteorological Society*, 94 (3), 329–343.

George, J.J., 1960: *Weather Forecasting for Aeronautics*. Academic Press, 673 pp.

Ghil, M., M. Halem, and R. Atlas, 1979: Time-continuous assimilation of remote-sounding data and its effect on weather forecasting. *Monthly Weather Review*, 107 (2), 140–171.

Godske, C.L., T. Bergeron, J. Bjerknes, and R.C. Bundgaard, 1957: *Dynamic Meteorology and Weather Forecasting*. American Meteorological Society, 800 pp.

Halem, M., M. Ghil, R. Atlas, J. Susskind, and W. Quirk, 1978: The GISS sounding temperature impact test. *NASA Tech. Mem. 78063*, 421 pp.

Halem, M., E. Kalnay, W.E. Baker, and R. Atlas, 1982: An assessment of the FGGE satellite observing systems during SOP-1. *Bulletin of the American Meteorological Society*, 63 (4), 407–426.

Halliwell, G.R., A. Srinivasan, H. Yang, D. Willey, M. Le Henaff, V. Kourafalou, and R. Atlas, 2014: Rigorous evaluation of a fraternal twin ocean OSSE system for the open Gulf of Mexico. *Journal of Oceanic and Atmospheric Technology*, 31 (1), 105–130.

Halliwell, G.R., V. Kourafalou, M. Le Henaff, L.K. Shay, and R. Atlas, 2015: OSSE impact analysis of airborne ocean surveys for improving upper-ocean dynamical and thermodynamical forces in the Gulf of Mexico. *Progress in Oceanography*, 130, 32–46.

Halliwell, G.R., M. Mehari, L.K. Shay, V.H. Kourafalou, H. Kang, H.-S. Kim, J. Dong, and R. Atlas, 2017a: OSSE quantitative assessment of rapid-response pre-storm ocean surveys to improve coupled tropical cyclone prediction. *Journal of Geophysical Research—Oceans*, 122 (7), 5729–5748.

Halliwell, G.R., M. Mehari, M. Le Henaff, V. Kourafalou, I. Androulidakis, H. Kang, and R. Atlas, 2017b: North Atlantic Ocean OSSE system: Evaluation of operational ocean observing system components and supplemental seasonal observations for improving coupled tropical cyclone prediction. *Journal of Operational Oceanography*, 10 (2), 154–175.

Hoffman, R.N., and R. Atlas, 2016a: Future observing system simulation experiments. *Bulletin of the American Meteorological Society*, 97 (9), 1601–1616.

Hoffman, R.N., and R. Atlas, 2016b: The OSSE checklist. *Bulletin of the American Meteorological Society*, 97 (9), ES193–ES199.

Kalnay, E., and R. Atlas, 1986: Global analysis of ocean surface wind and wind stress using the GLAS GCM and Seasat scatterometer winds. *Journal of Geophysical Research*, 91 (C2), 2233–2240.

Knudson, C., 1961: The New York Weather Bureau office in Rockefeller Center. *Weatherwise*, 43–49.

Kocin, P., and L. Uccellini, 2004: *Northeast Snowstorms*. American Meteorological Society, 818 pp.

Leidner, S.M., B. Annane, B. McNoldy, R. Hoffman, and R. Atlas, 2018: Variational analysis of simulated ocean surface winds from the Cyclone Global Navigation Satellite System (CYGNSS) and evaluation using a regional OSSE. *Journal of Atmospheric and Oceanic Technology*, 35 (8), 1571–1584.

Lenzen, A.J., D.R. Johnson, and R. Atlas, 1993: Analysis of the impact of Seasat scatterometer data and horizontal resolution on GLA model simulations of the QE II Storm. *Monthly Weather Review*, 121 (2), 499–521.

Lin, S.-J., 2004: A vertically Lagrangian finite-volume dynamical core for global models. *Monthly Weather Review*, 132, 2293–2307.

Malone, T.F., 1951: *Compendium of Meteorology*. American Meteorological Society, 1334 pp.

McNoldy, B., B. Annane, S. Majumdar, J. Delgado, L. Bucci, and R. Atlas, 2017: Impact of assimilating CYGNSS data on tropical cyclone analyses and forecasts in a regional OSSE framework. *Marine Technology Society Journal*, 51 (7), 7–15.

Mueller, M.J., A. Kren, L. Cucurull, R. Hoffman, R. Atlas, and T. Peevey, 2020: Impact of refractivity profiles from a proposed GNSS-RO constellation on tropical cyclone forecasts in a global modeling system. *Monthly Weather Review*, 148 (7), 3037–3057.

National Weather Analysis Center/Analysis and Forecast Branch, 1961: *Synoptic Meteorology as Practiced by the National Meteorological Center: The NAWAC Manual, Part II*. Government Printing Office, 37 pp.

Oke, P.R., G. Larnicol, E.M. Jones, V. Kourafalou, A.K. Sperrevik, F. Carse, C.A.S. Tanajura, B. Mourre, M. Tonani, G.B. Brassington, M. Le Hénaff, G.R. Halliwell, R. Atlas, A.M. Moore, C.A. Edwards, M.J. Martin, A.A. Sellar, A. Alvarez, P. De Mey, and M. Iskandarani, 2015: Assessing the impact of observations on ocean forecasts and reanalyses, Part 2—Regional applications. *Journal of Operational Oceanography*, 8 (S1), s63–s79.

Prive, N.C., Y. Xie, S. Koch, R. Atlas, S.J. Majumdar, and R.N. Hoffman, 2014: An observing system simulation experiment for the unmanned aircraft system data impact on tropical cyclone track forecasts. *Monthly Weather Review*, 142 (11), 4357–4363.

Pu, Z., L. Zhang, S. Zhang, B. Gentry, D. Emmitt, B. Demoz, and R. Atlas, 2017: The impact of Doppler wind lidar measurements on high-impact weather forecasting: Regional OSSE and data assimilation studies. In *Data Assimilation for Atmospheric, Oceanic and Hydrological Applications, Volume* 3, S.K. Park and L. Xu (eds.). Springer International, 259–283.

Rogers, R.F., 2021: Recent advances in our understanding of tropical cyclone intensity change processes from airborne observations. *Atmosphere*, 12 (5), 650, https://doi.org/10.3390/atmos12050650.

Ryan, K., L. Bucci, R. Atlas, J. Delgado, C. Landsea, and S. Murillo, 2019: Impact of GulfstreamIV dropsondes on tropical cyclone prediction in a regional OSSE system. *Monthly Weather Review*, 147 (8), 2961–2977.

Shen, B.-W., R. Atlas, J.-D. Chern, O. Reale, S.-J. Lin, T. Lee, and J. Chang, 2006a: The 0.125 degree finite-volume general circulation model on the NASA Columbia supercomputer: Preliminary simulations of mesoscale vortices. *Geophysical Research Letters,* 33 (5), L05801, https://doi.org/10.1029/2005GL024594.

Shen, B.-W., R. Atlas, O. Reale, S.-J. Lin, J.-D. Chern, J. Chang, C. Henze, and J.-L. Li, 2006b: Hurricane forecasts with a global mesoscale-resolving model: Preliminary results with Hurricane Katrina (2005). *Geophysical Research Letters,* 33 (13), L13813, https://doi.org/10.1029/2006GL026143.

Shen, B.-W., W.-K. Tao, W. K. Lau, and R. Atlas, 2010: Predicting tropical cyclogenesis prediction with a global mesoscale model: Hierarchical multiscale interactions during the formation of Tropical Cyclone Nargis (2008). *Journal of Geophysical Research,* 115, D14102, 15 pp.

Spar, J., and R. Atlas, 1975: Atmospheric response to sea surface temperature variations. *Journal of Applied Meteorology,* 14 (7), 1235–1245.

Spar, J., R. Atlas, and E. Kuo, 1976: Monthly mean forecast experiments with the GISS model. *Monthly Weather Review,* 104 (10), 1215–1241.

Susskind, J. and R. M. Atlas, 2004: Atmospheric soundings from AIRS/AMSU/HSB. *Algorithms and Technologies for Multispectral, Hyperspectral, and Ultraspectral Imagery X, Proceedings, SPIE,* Vol. 5425, 311–319.

Tracton, M. S., A. J. Desmaris, R. D. McPherson, and R. J. Van Haaren, 1980: The impact of satellite soundings upon the National Meteorological Centers analysis and forecast system-The Data System tests results. *Monthly Weather Review,* 108, 543–586.

Tracton, M. S., A. J. Desmaris, R. D. McPherson, and R. J. Van Haaren, 1981: On the system dependency of satellite sounding impact—comments on recent impact test results. *Monthly Weather Review,* 109, 197–200.

Uccellini, L., R. A. Petersen, C. H. Wash, and K. F. Brill, 1984: The President's Day cyclone of 18–19 February 1979: Synoptic overview and analysis of the subtropical jet streak influencing the precyclogenetic period. *Monthly Weather Review,* 112, 31–55.

Uccellini, L., R. A. Petersen, K. F. Brill, P. J. Kocin, and J.J. Tuccillo,1987: Synergistic interactions between an upper-level jet streak and diabatic processes that influence the development of a low-level jet and a secondary coastal cyclone. *Monthly Weather Review,* 115, 2227–2261.

Vergin, J.M., D.R. Johnson, and R. Atlas, 1984: A quasi-Lagrangian diagnostic case study of the effect of satellite sounding data assimilation on model cyclone prediction. *Monthly Weather Review,* 112 (4), 725–739.

Wolfson, N., R. Atlas, and Y. Sud, 1985: A numerical investigation of the summer 1980 heat wave and drought. *Proceedings, 7th AMÍ Conference on Numerical Weather Prediction, Montreal, Canada, June 17–20,* 267–272.

Wolfson, N., and R. Atlas, 1986: A simple diagnostic tool for the investigation of persistent phenomena with application to the summer 1980 heat wave. *Atmosphere-Ocean,* 24 (2), 111–127.

Wolfson, N., R. Atlas, and Y.C. Sud, 1987: Numerical experiments related to the summer 1980 U.S. heat wave. *Monthly Weather Review*, 115 (7), 1345–1357.

Zhang, J.A., R. Atlas, G.D. Emmitt, L. Bucci, and K. Ryan, 2018: Airborne Doppler wind lidar observations of the tropical cyclone boundary layer. *Remote Sensing*, 10 (6), 825.

Zhang, S., Z. Pu, D. Posselt, and R. Atlas, 2017: An impact of CYGNSS ocean surface wind speeds on numerical simulations of a hurricane in observing system simulation experiments. *Journal of Oceanic and Atmospheric Technology*, 34 (2), 375–383.

Zheng, X., R. Atlas, R.J. Birk, F.H. Carr, M.J. Carrier, L. Cucurull, W.H. Hooke, E. Kalnay, R. Murtugudde, D.J. Posselt, J.L. Russell, D.P. Tyndall, R.A. Weller, and F. Zhang, 2020: Use of observing system simulation experiments in the U.S. *Bulletin of the American Meteorological Society*, 101 (8), E1427–E1438.

Appendix A: Career Timeline

Education

1970	BS, Aeronautics, Parks College of Aeronautical Technology of St. Louis University
1973	MS, Meteorology, New York University
1976	PhD, Meteorology and Oceanography, New York University

Previous Positions

1961–1966	Apprentice Weather Forecaster, U.S. Weather Bureau, New York
1970	Research Assistant, Parks College Aeronautical Studies Group
1970–1972	Weather Forecaster-Briefer, USAF Air Weather Service
1973	Summer Intern, National Center for Atmospheric Research
1973	Part-time Instructor of Physics, State University of New York, Agricultural and Technical College
1973–1976	Research Assistant, City University Research Foundation
1974–1977	Adjunct Instructor of Meteorology and Oceanography, State University of New York Maritime College
1975	Chief Consulting Meteorologist of *AM America* Show, American Broadcasting Company
1975–1978	Assistant Professor of Atmospheric and Oceanic Science, State University of New York at Stony Brook
1976–1977	Research Associate, NASA/Goddard Institute for Space Studies
1978–1998	Research Scientist, NASA/Goddard Space Flight Center and Adjunct Assistant Professor, University of Maryland

1998–2003	Head, NASA Data Assimilation Office and Adjunct Professor, University of Maryland
2003–2005	Chief Meteorologist, Laboratory for Atmospheres, NASA Goddard Space Flight Center
2005–2019	Director, NOAA Atlantic Oceanographic and Meteorological Laboratory and Adjunct Professor, University of Miami
2014–2019	Director, NOAA's Quantitative Observing System Assessment Program
2019–present	Director Emeritus, Atlantic Oceanographic and Meteorological Laboratory

Past and Current Professional Society Memberships

- Fellow and former Councilor, American Meteorological Society
- Chi Epsilon Pi, Meteorology Honor Society
- American Association for the Advancement of Science
- American Geophysical Union
- International Society for Optical Engineering
- Charter Member of National Weather Association
- Past member of Forecasting and Education and Training Committees of the National Weather Association
- Former President of the Stony Brook Chapter of the American Meteorological Society

Awards/Honors

1971	Honor Graduate of Advanced Weather Techniques Course at Chanute Technical Training Center, USAF
1976	National Research Council Research Associateship
1979	NASA Special Achievement Award
1979	NASA Certificates for Outstanding Performance,
1979	Laboratory for Atmospheric Science—Outstanding Paper Award
1982	NASA Special Achievement Award
1982	Meritorious Service Award, Global Modeling and Simulation Branch
1984	NASA Special Achievement Awards
1985	Science Achievement Award, Laboratory for Atmospheres
1992	Scientific Research Award, Laboratory for Atmospheres
1995	Invited Lead-off Speaker for WMO International Symposium on Data Assimilation
1998	NASA Group Achievement Award
2001	NASA Medal for Exceptional Scientific Achievement
2007	NASA Group Achievement Award
2010	NASA Group Achievement Award

2016 American Meteorological Society Banner I. Miller Award
2019 Honored on the floor of the U.S. House of Representatives
2019 Honored by National Hurricane Center for enduring contributions to the nation's hurricane forecast and warning program
2019 Invited Keynote Speaker for the International Winter Simulation Conference

Special Experience

- Project scientist for FGGE Special Effort for satellite data enhancement
- Member of NASA Satellite Surface Stress Committee
- Member of NASA Scatterometer (NSCAT) Science Team
- Member of Laser Atmospheric Wind Sounder (LAWS) Science Team
- Member of SeaWinds Instrument Teams
- Member of Ocean Vector Winds Science Team
- Chairman of U.S. WOCE Advisory Panel for Model-based Air Sea Fluxes
- Member of NASA/NOAA Working Group for Space-Based LIDAR Winds
- Team Leader for User Requirements for AEOLIS/WINDRAD
- Zephyr Wind Lidar Project Scientist
- Member of North American Observing System (NAOS) Test and Evaluation Working Group
- Member of the Scientific Steering Group for GEWEX
- Member of USWRP Scientific Steering Committee
- Member of Thorpex Scientific Steering Committee
- Member of Management Oversight Board for the Joint Center for Satellite Data Assimilation (JCSDA)
- Member of Executive Oversight Board for the Hurricane Forecast Improvement Project (HFIP)
- OAR Representative to NOAA Research Council
- Program Chair for AMS Symposium on Observations, Data Assimilation and Probabilistic Prediction (2002)
- Program Chair for AMS Integrated Observing and Assimilation Systems for the Atmosphere, Ocean, and Land Surface (IOAS-AOLS) Conference (2006–2021)

Appendix B: Publications of Robert M. Atlas

Refereed Publications

1975–79

Spar, J., and R. Atlas, 1975: Atmospheric response to sea surface temperature variations. *Journal of Applied Meteorology,* 14 (7), 1235–1245.

Spar, J., R. Atlas, and E. Kuo, 1976: Monthly mean forecast experiments with the GISS model. *Monthly Weather Review*, 104 (10), 1215–1241.

Halem, M., M. Ghil, and R. Atlas, 1978: Some experiments on the effect of remote sounding temperatures upon weather forecasting. In *Remote Sensing of the Atmospheres: Inversion Methods and Applications*. Elsevier, Amsterdam, 9–33.

Ghil, M., M. Halem, and R. Atlas, 1979: Effects of sounding temperature assimilation on weather forecasting: Model dependence studies. In *Remote Sounding of the Atmosphere from Space*. Pergamon Press, Oxford and New York, 21–25.

Ghil, M., M. Halem, and R. Atlas, 1979: Time-continuous assimilation of remote-sounding data and its effect on weather forecasting. *Monthly Weather Review*, 107 (2), 140–171.

1980–89

Atlas, R., M. Ghil, and M. Halem, 1981: Reply to comment by L. Druyan on time-continuous assimilation of remote-sounding data and its effect on weather forecasting. *Monthly Weather Review*, 109 (1), 201–204.

Atlas, R., 1982: The growth of prognostic differences between GLAS model forecasts from SAT and NOSAT initial condition. *Monthly Weather Review*, 110 (7), 877–882.

Atlas, R., M. Halem, and M. Ghil, 1982: The effect of model resolution and satellite sounding data on GLAS model forecasts. *Monthly Weather Review*, 110 (7), 662–682.

Halem, M., E. Kalnay, W.E. Baker, and R. Atlas, 1982: An assessment of the FGGE satellite observing systems during SOP-1. *Bulletin of the American Meteorological Society*, 63 (4), 407–426.

Atlas, R., E. Kalnay, and M. Halem, 1984: Review of experiments on the impact of satellite data on numerical weather prediction. *SPIE*, Vol. 481, 108–115.

Baker, W.E., R. Atlas, M. Halem, and J. Susskind, 1984: A case study of forecast sensitivity to data and data analysis techniques. *Monthly Weather Review*, 112 (8), 1544–1561.

Baker, W.E., R. Atlas, E. Kalnay, M. Halem, P.M. Woiceshyn, and D. Edelmann, 1984: Large-scale analysis and forecast experiments with wind data from the Seasat-A scatterometer. *Journal of Geophysical Research*, 89, 4927–4936.

Duffy, D., R. Atlas, T. Rosmond, E. Barker, and R. Rosenberg, 1984: The impact of Seasat scatterometer winds on the Navy's operational model. *Journal of Geophysical Research*, 89, 7238–7244.

Vergin, J.M., D.R. Johnson, and R. Atlas, 1984: A quasi-Lagrangian diagnostic case study of the effect of satellite sounding data assimilation on model cyclone prediction. *Monthly Weather Review*, 112 (4), 725–739.

Atlas, R., E. Kalnay, and M. Halem, 1985: Impact of satellite temperature sounding and wind data on numerical weather prediction. *Optical Engineering*, 24 (2), 341–346.

Duffy, D., and R. Atlas, 1986: The impact of Seasat-A scatterometer data on the numerical prediction of the QE II storm. *Journal of Geophysical Research*, 91 (C2), 2241–2248.

Kalnay, E., and R. Atlas, 1986: Global analysis of ocean surface wind and wind stress using the GLAS GCM and Seasat scatterometer winds. *Journal of Geophysical Research*, 91 (C2), 2233–2240.

Wolfson, N., and R. Atlas, 1986: A simple diagnostic tool for the investigation of persistent phenomena with application to the summer 1980 heat wave. *Atmosphere-Ocean*, 24 (2), 111–127.

Atlas, R., 1987: The role of oceanic fluxes and initial data in the numerical prediction of an intense coastal storm. *Dynamics of Atmospheres and Oceans*, 10 (4), 359–388.

Atlas, R., A.J. Busalacchi, E. Kalnay, S. Bloom, and M. Ghil, 1987: Global surface wind and flux fields from model assimilation of Seasat data. *Journal of Geophysical Research*, 92 (C6), 6477–6487.

Helfand, H.M., J. Pfaendtner, and R. Atlas, 1987: The effect of increased horizontal resolution on GLA fourth order model forecasts. *Journal of the Meteorological Society of Japan*, 303–315.

Wolfson, N., R. Atlas, and Y.C. Sud, 1987: Numerical experiments related to the summer 1980 U.S. heat wave. *Monthly Weather Review*, 115 (7), 1345–1357.

1990–99

Atlas, R., S.C. Bloom, R.N. Hoffman, J.V. Ardizzone, and G. Brin, 1991: Space-based surface wind vectors to aid understanding of air-sea interactions. *EOS, Transactions, American Geophysical Union*, 72, 201–208.

Jusem, J.C., and R. Atlas, 1991: Diagnostic evaluation of numerical model simulations using the tendency equation. *Monthly Weather Review*, 119 (12), 2936–2955.

Atlas, R., R.W. Hoffman, and S.C. Bloom, 1993: Surface wind velocity over the oceans. *Atlas of Satellite Observations Related to Global Change Atlas*, 129–139.

Atlas, R., N. Wolfson, and J. Terry, 1993: The effect of SST and soil moisture anomalies on GLA model simulations of the 1988 U.S. summer drought. *Journal of Climate*, 6 (11), 2034–2048.

Busalacchi, A.J., R. Atlas, and E.C. Hackert, 1993: Comparison of special sensor microwave imager vector wind stress with model-derived and subjective products for the tropical Pacific. *Journal of Geophysical Research*, 98 (C4), 6961–6978.

Lenzen, A.J., D.R. Johnson, and R. Atlas, 1993: Analysis of the impact of Seasat scatterometer data and horizontal resolution on GLA model simulations of the QE II Storm. *Monthly Weather Review*, 121 (2), 499–521.

Liu, W.T., W. Tang, and R. Atlas, 1993: Sea surface temperature exhibited by an ocean general circulation model in response to wind forcing derived from satellite data. In *Remote Sensing of the Oceanic Environment*. Seibutso Kenkyusha, Tokyo, Japan, 350–355.

Atlas, R., 1995: Atmospheric observations. *Geophysical Magazine*, 2, 1–18.

Baker, W.E., G.D. Emmitt, F. Robertson, R.M. Atlas, J.E. Molinari, D.A. Bowdle, J. Paegle, R.M. Hardesty, R.T. Menzies, T.N. Krishnamurti, R.A. Brown, M.J. Post, J.R. Anderson, A.C. Lorenc, and J. McElroy, 1995: Lidar measured winds from space: A key component for weather and climate prediction. *Bulletin of the American Meteorological Society*, 76 (6), 869–888.

Chou, S.H., R.M. Atlas, C.-L. Shie, and J. Ardizonne, 1995: Estimates of surface humidity and latent heat fluxes over oceans from SSM/I data. *Monthly Weather Review*, 123 (8), 2405–2425.

Walker, G.K., Y.C. Sud, and R. Atlas, 1995: Impact of the ongoing Amazonian deforestation on local precipitation: A GCM simulation study. *Bulletin of the American Meteorological Society*, 76 (3), 346–361.

Atlas, R., R.N. Hoffman, S.C. Bloom, J.C. Jusem, and J. Ardizzone, 1996: A multiyear global surface wind velocity dataset using SSM/I wind observations. *Bulletin of the American Meteorological Society*, 77 (5), 869–882.

Liu, W.T., W. Tang, and R. Atlas, 1996: Responses of the tropical Pacific to wind forcing as observed by spaceborne sensors and simulated by an ocean general circulation model. *Journal of Geophysical Research*, 101 (C7), 16,345–16,360.

Rienecker, M.M., R. Atlas, S.D. Schubert, and C.S. Willett, 1996: A comparison of wind products over the North Pacific Ocean. *Journal of Geophysical Research*, 101 (C1), 1011–1024.

Atlas, R., 1997: Atmospheric observations and experiments to assess their usefulness in data assimilation. *Journal of the Meteorological Society of Japan*, 75 (1B), 111–130.

Chou, S.-H., C.-L. Shie, R.M. Atlas, and J. Ardizzone, 1997: Air-sea fluxes retrieved from special sensor microwave imager data. *Journal of Geophysical Research,* 102 (C6), 12,705–12,726.

Chin, M., R.B. Rood, D.J. Allen, M.O. Andreae, A.M. Thompson, S.-J. Lin, R.M. Atlas, and J.V. Ardizzone, 1998: Processes controlling dimethylsulfide over the ocean: Case studies using a 3-D model driven by assimilated meteorological fields. *Journal of Geophysical Research,* 103 (D7), 8341–8354.

Jusem, J.C., and R. Atlas, 1998: Diagnostic evaluation of vertical motion forcing mechanisms by using Q-vector partitioning. *Monthly Weather Review,* 126 (8), 2166–2184.

Atlas, R., 1999: Wind speed and velocity. In *Remote Sensing Data Book.* Cambridge University Press.

Atlas, R., S.C. Bloom, R.N. Hoffman, E. Brin, J. Ardizzone, J. Terry, D. Bungato, and J.C. Jusem, 1999: Geophysical validation of NSCAT winds using atmospheric data and analyses. *Journal of Geophysical Research,* 104 (C5), 11,405–11,424.

Otterman, J., R. Atlas, J. Ardizzone, D. Starr, J.C. Jusem, and J. Terry, 1999: Relationship of late-winter temperatures in Europe to North Atlantic surface winds: A correlation analysis. *Journal of Theoretical and Applied Climatology,* 64 (3–4), 201–211.

Otterman, J., R. Atlas, D. Starr, J. Ardizzone, and J. Terry, 1999: Space observations of ocean surface winds aid monitoring of northeast Pacific climate shifts. *EOS, Transactions, American Geophysical Union,* 79, 575–581.

2000–2009

Atlas, R., and R.N. Hoffman, 2000: The use of satellite surface wind data to improve weather analysis and forecasting at the NASA Data Assimilation Office. In *Satellites, Oceanography, and Society,* D. Halpern (ed.). Elsevier Oceanography Series, 63, 1–22.

Hou, A.Y., D.V. Ledvina, A.M. da Silva, S.Q. Zhang, J. Joiner, R.M. Atlas, G.J. Huffman, and C.D. Kummerow, 2000: Assimilation of SSM/I-derived surface rainfall and total precipitable water for improving the GEOS analysis for climate studies. *Monthly Weather Review,* 128 (3), 509–537.

Otterman, J., J. Ardizzone, R. Atlas, J.C. Jusem, A. Karnieli, H. Saaroni, D. Starr, and J. Terry, 2000: Application of the special sensor microwave imager dataset to observing circulation in the Mediterranean basin. *International Journal of Remote Sensing.*

Otterman, J., J. Ardizzone, R. Atlas, H. Hu, J.C Jusem, and D. Starr, 2000: Winter-to-spring transition in Europe 48–54°N: From temperature control by advection to control by isolation. *Geophysical Research Letters,* 27 (4), 561–564.

Atlas, R., R.N. Hoffman, S.M. Leidner, J. Sienkiewicz, T.-W. Yu, S.C. Bloom, E. Brin, J. Ardizzone, J. Terry, D. Bungato, and J.C. Jusem, 2001: The effects of marine winds from scatterometer data on weather analysis and forecasting. *Bulletin of the American Meteorological Society,* 82 (9), 1965–1990.

Conaty, A.L., J.C. Jusem, L. Takacs, D. Keyser, and R. Atlas, 2001: The structure and evolution of extratropical cyclones, fronts, jet streams, and the tropopause in the GEOS general circulation model. *Bulletin of the American Meteorological Society,* 82 (9), 1853–1867.

Otterman, J., J.K. Angell, R. Atlas, J. Ardizzone, G. Demarée, J.C. Jusem, D. Koslowsky, and J. Terry, 2001: The extremely warm early winter 2000 in Europe: What is the forcing? *EOS, Transactions, American Geophysical Union.*

Otterman, J., J. Ardizzone, R. Atlas, D. Bungato, J. Cierniewski, J.C. Jusem, R. Przybylak, S. Schubert, D. Starr, J. Walczewski, and A. Wos, 2001: Extreme anomalies of winter air temperature in mid-latitude Europe. *Geographia Polonica,* 74 (2), 57–67.

Reale, O., and R. Atlas, 2001: Tropical cyclone-like vortices in the extratropics: Observational evidence and synoptic analysis. *Weather and Forecasting,* 16 (1), 7–34.

Atlas, R., A. Hou, and O. Reale, 2002: Hurricane and flood prediction, community disaster preparedness. *Earth Observation Magazine,* 11 (8).

Milliff, R.F., M.H. Freilich, W.T. Liu, R. Atlas, and W.G. Large, 2002: Global ocean surface vector wind observations from space. In *Observing the Ocean in the 21st Century from Space,* C.J. Koblinsky and N.R. Smith (eds.). GODAE Project Office and Beareau of Meteorology, Melbourne, 102–119.

Otterman, J., J.K. Angell, J. Ardizzone, R. Atlas, S. Schubert, D. Starr, and M.-L.C. Wu, 2002: North-Atlantic surface winds examined as the source of winter warming in Europe. *Geophysical Research Letters,* 29 (19), https://doi.org/10.1029/2002GL015256.

Otterman, J., J.K. Angell, R. Atlas, D. Bungato, S. Schubert, D. Starr, J. Susskind, and M.-L.C. Wu, 2002: Advection from the North Atlantic as the forcing of winter greenhouse effect over Europe. *Geophysical Research Letters,* 29 (8), https://doi.org/10.1029/2001GL014187.

Otterman, J., R. Atlas, S.-H. Chou, J.C. Jusem, R.A. Pielke, T.N. Chase, J. Rogers, G.L. Russell, S.D. Schubert, Y.C. Sud, and J. Terry, 2002: Are stronger North-Atlantic southwesterlies the forcing to the late-winter warming in Europe? *International Journal of Climatology,* 22 (6), 743–750.

Chou, S.-H., E. Nelkin, J. Ardizzone, R. Atlas, and C.-L. Shie, 2003: Surface turbulent heat and momentum fluxes over global oceans on basis of the Goddard satellite retrievals, Version 2 (GSSTF2). *Journal of Climate,* 16 (20), 3256–3273.

Fetzer, E., L.M. McMillin, D. Tobin, H.H. Aumann, M.R. Gunson, W.W. McMillan, D.E. Hagan, M.D. Hofstadter, J. Yoe, D.N. Whiteman, J.E. Barnes, R. Bennartz, H. Vomel, V. Walden, M. Newchurch, P.J. Minnett, R. Atlas, F. Schmidlin, E.T. Olsen, M.D. Goldberg, S. Zhou, H.-J. Ding, W.L. Smith, and H. Revercomb, 2003: AIRS/AMSU/HSB validation. *IEEE Transactions on Geoscience and Remote Sensing,* 41 (2), 418–431.

Henderson, J.M., R.N. Hoffman, S.M. Leidner, R. Atlas, E. Brin, and J.V. Ardizzone, 2003: A comparison of a two-dimensional variational analysis method and a median filter for NSCAT ambiguity removal. *Journal of Geophysical Research,* 108 (C6), https://doi.org/10.1029/2002JC001307.

Hoffman, R.N., S.M. Leidner, J.M. Henderson, R. Atlas, J.V. Ardizzone, and S.C. Bloom, 2003: A two-dimensional variational analysis method for NSCAT ambiguity removal: Methodology, sensitivity, and tuning. *Journal of Atmospheric and Oceanic Technology,* 20 (5), 585–605.

Otterman, J., R. Atlas, G.L. Russell, and H. Saaroni, 2003: Impact on regional winter climate by CO_2 increases vs. by maritime-air advection. *Geophysical Research Letters,* 30 (15), https://doi.org/10.1029/ 2003GL018587.

Otterman, J., D.O'C. Starr, R. Atlas, J.C. Jusem, and H. Saaroni, 2003: Circumpolar circulation patterns over the Northern Hemisphere oceans in late winter, 1949–2002. *Meter. Zeit.*

Sud, Y.C., D.M. Mocko, K.-M. Lau, and R. Atlas, 2003: On simulating the midwestern U.S. drought of 1988 with a GCM. *Journal of Climate,* 16 (23), 3946–3965.

Chou, S.-H., E. Nelkin, J. Ardizzone, and R. Atlas, 2004: A comparison of latent heat fluxes over global oceans for four flux products. *Journal of Climate,* 17 (20), 3973–3989.

Crisp, D., R. Atlas, F.-M. Breon, L.R. Brown, J.P. Burrows, P. Ciais, B.J. Connor, S.C. Doney, I.Y. Fung, D.J. Jacob, C.E. Miller, D. O'Brien, S. Pawson, J.T. Randerson, P. Rayner, R.J. Salawitch, S.P. Sander, B. Sen, G.L. Stephens, P.P. Tans, G.C. Toon, P.O. Wennberg, S.C. Wofsy, Y.L. Yung, Z. Kuang, B. Chudasama, G. Sprague, B. Weiss, R. Pollock, D. Kenyon, and S. Schroll, 2004: The Orbiting Carbon Observatory (OMO) mission. *Advances in Space Research,* 34 (4), 700–709.

Lin, S.-J., R. Atlas, and K.-S. Yeh, 2004: Global weather prediction and high-end computing at NASA. *Computing in Science and Engineering,* 6 (1), 29–34.

Riishojgaard, L.P., R. Atlas, and G.D. Emmitt, 2004: The impact of Doppler lidar wind observations on a single-level meteorological analysis. *Journal of Applied Meteorology,* 43 (5), 810–820.

Atlas, R., A.Y. Hou, and R. Oreste, 2005: Application of SeaWinds scatterometer and TMI-SSM/I rain rates to hurricane analysis and forecasting. *Journal of Photogrammetry and Remote Sensing,* 59 (4), 233–243.

Atlas, R., R.O. Reale, B.-W. Shen, S.-J. Lin, J.-D. Chern, W. Putman, T. Lee, K.-S. Yeh, M. Bosilovich, and J. Radakovich, 2005: Hurricane forecasting with the high-resolution NASA finite volume general circulation model. *Geophysical Research Letters,* 32 (3), L03807, https://doi.org/10.1029/2004GL021513.

Li, J.-L., D.E. Waliser, J.H. Jiang, D.L. Wu, W. Read, J.W. Waters, A.M. Tompkins, L.J. Donner, J.-D. Chern, W.-K. Tao, R. Atlas, Y. Gu, K.N. Liou, A. Del Genio, M. Khairoutdinov, and A. Gettelman, 2005: Comparisons of EOS MLS cloud ice measurements with ECMWF analyses and GCM simulations: Initial results. *Geophysical Research Letters,* 32 (18), L18710, https://doi.org/10.1029/2005GL023788.

Chahine, M.T., T.S. Pagano, H.H. Aumann, R. Atlas, C. Barnet, J. Blaisdell, L. Chen, M. Divakarla, E.J. Fetzer, M. Goldberg, C. Gautier, S. Granger, S. Hannon, F.W. Irion,

R. Kakar, E. Kalnay, B.H. Lambrigtsen, S.-Y. Lee, J. LeMarshall, W.W. McMillan, L. McMillin, E.T. Olsen, H. Revercomb, P. Rosenkranz,W.L. Smith, D. Staelin, L.L. Strow, J. Susskind, D. Tobin, W. Wolf, and L. Zhou, 2006: AIRS: Improving weather forecasting and providing new data on greenhouse gases. *Bulletin of the American Meteorological Society,* 87 (7), 911–926.

Riishojgaard, L.P., R. Atlas, and G.D. Emmitt, 2006: Reply to "The impact of Doppler lidar wind observations on a single-level meteorological analysis." *Journal of Applied Meteorology and Climatology,* 45 (6), 887–888.

Shen, B.-W., R. Atlas, J.-D. Chern, O. Reale, S.-J. Lin, T. Lee, and J. Chang, 2006: The 0.125 degree finite-volume general circulation model on the NASA Columbia supercomputer: Preliminary simulations of mesoscale vortices. *Geophysical Research Letters,* 33 (5), L05801, https://doi.org/10.1029/2005GL024594.

Shen, B.-W., R. Atlas, O. Reale, S.-J. Lin, J.-D. Chern, J. Chang, C. Henze, and J.-L. Li, 2006: Hurricane forecasts with a global mesoscale-resolving model: Preliminary results with Hurricane Katrina (2005). *Geophysical Research Letters,* 33 (13), L13813, https://doi.org/10.1029/2006GL026143.

Atlas, R., S.-J. Lin, B.-W. Shen, O. Reale, and K.-S. Yeh, 2007: Improving hurricane prediction through innovative global modeling. In *Extending the Horizons: Advances in Computing, Optimization, and Decision Technologies,* E.K. Baker, A. Joseph, A. Mehrotra, and M.A. Trick (eds.). Springer, 1–14.

Joiner, J., E. Brin, R. Treadon, J. Derber, P. Van Delst, A. Da Silva, J. Le Marshall, P. Poli, R. Atlas, D. Bungato, and C. Cruz, 2007: Effects of data selection and error specification on the assimilation of AIRS data. *Quarterly Journal of the Royal Meteorological Society,* 133 (622), 181–196.

Atlas, R., and G.D. Emmitt, 2008: Review of observing system simulation experiments to evaluate the potential impact of lidar winds on numerical weather prediction. *ILRC24,* Vol. 2 (ISBN 978-0-615-21489-4), 726–729.

Gentry, B., M. McGill, G. Schwemmer, M. Hardesty, A. Brewer, T. Wilkerson, R. Atlas, M. Sirota, S. Lindemann, and F. Hovis, 2008: New technologies for direct detection of Doppler lidar: Status of the TWiLiTE airborne molecular Doppler lidar project. *ILRC24,* Vol. 1 (ISBN 978-0-615-21489-4), 239–243.

Ardizonne, J., R. Atlas, R.N. Hoffman, J.C. Jusem, S.M. Leidner, and D.F. Moroni, 2009: New multiplatform ocean surface wind product available. *EOS, Transactions, American Geophysical Union,* 90 (27), 231.

Tao, W.-K., J.-D. Chern, R. Atlas, D. Randall, M. Khairoutdinov, J.-L. Li, D.E. Waliser, A. Hou, X. Lin, C. Peters-Lidard, W. Lau, J. Jiang, and J. Simpson, 2009: A multiscale modeling system: Developments, applications, and critical issues. *Bulletin of the American Meteorological Society,* 9 (4), 515–534.

2010–19

Shen, B.-W., W.-K. Tao, W. K. Lau, and R. Atlas, 2010: Predicting tropical cyclogenesis prediction with a global mesoscale model: Hierarchical multiscale interactions during the formation of Tropical Cyclone Nargis (2008). *Journal of Geophysical Research,* 115, D14102, 15 pp.

Atlas, R., R.N. Hoffman, J. Ardizzone, S.M. Leidner, J.C. Jusem, D.K. Smith, and D. Gombos, 2011: A cross-calibrated, multi-platform ocean surface wind velocity product for meteorological and oceanographic applications. *Bulletin of the American Meteorological Society,* 92 (2), 157–174.

Chen, H., D.-L. Zhang, J. Carton, and R. Atlas, 2011: On the rapid intensification of Hurricane Wilma (2005). Part I: Model prediction and structural changes. *Weather and Forecasting,* 26 (6), 885–901.

Gopalakrishnan, S.G., F. Marks, X. Zhang, J.-W. Bao, K.-S. Yeh, and R. Atlas, 2011: The Experimental HWRF system: A study of the influence of horizontal resolution on the structure and intensity changes in tropical cyclones using an idealized framework. *Monthly Weather Review,* 139 (6), 1762–1784.

Wang, C., H. Liu, S.-K. Lee, and R. Atlas, 2011: Impact of the Atlantic warm pool on United States landfalling hurricanes. *Geophysical Research Letters,* 38, L19702, 7 pp.

Gopalakrishnan, S.G., S. Goldenberg, T. Quirino, F. Marks, X. Zhang, K.-S. Yeh, R. Atlas, and V. Tallapragada, 2012: Towards improving high-resolution numerical hurricane forecasting: Influence of model horizontal grid resolution, initialization, and physics. *Weather and Forecasting,* 27 (3), 647–666.

Yeh, K.-S., X. Zhang, S.G. Gopalakrishnan, S. Aberson, R. Rogers, F.D. Marks, and R. Atlas, 2012: Performance of the experimental HWRF in the 2008 hurricane season. *Natural Hazards,* 63 (3), 1439–1449.

Hoffman, R.N., J.V. Ardizzone, S.M. Leidner, D.K. Smith, J.C. Jusem, and R.M. Atlas, 2013: Error estimates for ocean surface winds: Applying Desroziers diagnostics to the cross-calibrated, multi-platform analysis of wind speed. *Journal of Oceanic and Atmospheric Technology,* 30 (11), 2596–2603.

Lee, S.-K., D.B. Enfield, H. Liu, C. Wang, and R. Atlas, 2013: Is there an optimal ENSO pattern that enhances large-scale atmospheric processes conducive to major tornado outbreaks in the United States? *Journal of Climate,* 26 (5), 1626–1642.

Nolan, D.S., R. Atlas, K.T. Bhatia, and L.R. Bucci, 2013: Development and validation of a hurricane nature run using the joint OSSE nature run and the WRF model. *Journal of Advances in Modeling Earth Systems,* 5 (2), 382–405.

Prive N.C., Y. Xie, J. Woollen, S. Koch, R. Atlas, and R. Hood, 2013: Evaluation of the Earth Systems Research Laboratory (ESRL) global Observing System Simulation Experiment (OSSE) system. *Tellus A,* 65, 19011, 22 pp.

Ralph, F.M., J. Intrieri, D. Andra, R. Atlas, S. Boukabara, D. Bright, P. Davidson, B. Entwistle, J. Gaynor, S. Goodman, J.-G. Jiing, A. Harless, J. Huang, G. Jedlovec, J. Kain, S. Koch, B. Kuo, J. Levit, S. Murillo, L.P. Riishojgaard, T. Schneider, R. Schneider, T. Smith, and S. Weiss, 2013: The emergence of weather-related testbeds linking research and forecasting operations. *Bulletin of the American Meteorological Society,* 94 (8), 1187–1211.

Baker, W.E., R. Atlas, C. Cardinali, A. Clement, G.D. Emmitt, B.M. Gentry, R.M. Hardesty, E. Kallen, M.J. Kavaya, R. Langland, M. Masutani, W. McCarty, R.B. Pierce, Z. Pu, L.P. Riishojgaard, J. Ryan, S. Tucker, M. Weissmann, and J.G. Yoe, 2014: Lidar-measured wind profiles: The missing link in the global observing system. *Bulletin of the American Meteorological Society,* 95 (4), 543–564.

Halliwell, G.R., A. Srinivasan, H. Yang, D. Willey, M. Le Henaff, V. Kourafalou, and R. Atlas, 2014: Rigorous evaluation of a fraternal twin ocean OSSE system for the open Gulf of Mexico. *Journal of Oceanic and Atmospheric Technology,* 31 (1), 105–130.

Prive, N.C., Y. Xie, S. Koch, R. Atlas, S.J. Majumdar, and R.N. Hoffman, 2014: An observing system simulation experiment for the unmanned aircraft system data impact on tropical cyclone track forecasts. *Monthly Weather Review,* 142 (11), 4357–4363.

Atlas, R., V. Tallapragada, and S. Gopalakrishnan, 2015: Advances in tropical cyclone intensity forecasts. *Marine Technology Society Journal,* 49 (6), 149–160.

Atlas, R., L. Bucci, B. Annane, R. Hoffman, and S. Murillo, 2015: Observing System Simulation Experiments to assess the potential impact of new observing systems on hurricane forecasting. *Marine Technology Society Journal,* 49 (6), 140–148.

Atlas, R., R.N. Hoffman, Z. Ma, G.D. Emmitt, S.A. Wood, S. Greco, S. Tucker, L. Bucci, B. Annane, and S. Murillo, 2015: Observing system simulation experiments (OSSEs) to evaluate the potential impact of an optical autocovariance wind lidar (OAWL) on numerical weather prediction. *Journal of Atmospheric and Oceanic Technology,* 32 (9), 1593–1613.

Goldenberg, S.B., S.G. Gopalakrishnan, T. Quirino, F. Marks, V. Tallapragada, S. Trahan, X. Zhang, and R. Atlas, 2015: The 2012 triply-nested, high-resolution operational version of the Hurricane Weather Research and Forecasting System (HWRF): Track and intensity forecast verifications. *Weather and Forecasting,* 30 (3), 710–729.

Halliwell, G.R., V. Kourafalou, M. Le Henaff, L.K. Shay, and R. Atlas, 2015: OSSE impact analysis of airborne ocean surveys for improving upper-ocean dynamical and thermodynamical forces in the Gulf of Mexico. *Progress in Oceanography,* 130, 32–46.

Oke, P.R., G. Larnicol, E.M. Jones, V. Kourafalou, A.K. Sperrevik, F. Carse, C.A.S. Tanajura, B. Mourre, M. Tonani, G.B. Brassington, M. Le Hénaff, G.R. Halliwell, R. Atlas, A.M. Moore, C.A. Edwards, M.J. Martin, A.A. Sellar, A. Alvarez, P. De Mey, and M. Iskandarani, 2015: Assessing the impact of observations on ocean forecasts and reanalyses, Part 2—Regional applications. *Journal of Operational Oceanography,* 8 (S1), s63–s79.

Androulidakis, Y.S., V.H. Kourafalou, G.R. Halliwell, M. Le Henaff, H.S. Kang, M. Mehari, and R. Atlas, 2016: Hurricane interaction with the upper ocean in the Amazon-Orinoco plume region. *Ocean Dynamics,* 66 (12), 1559–1588.

Boukabara, S.A., I. Moradi, R. Atlas, S.P.F. Casey, L. Cucurull, R.N. Hoffman, K. Ide, V. Krishna Kumar, R. Li, Z. Li, M. Masutani, N. Shahroudi, J. Woollen, and Y. Zhou, 2016: Community global Observing System Simulation Experiment (OSSE) package: CGOP—Description and usage. *Journal of Atmospheric and Oceanic Technology,* 33 (8), 1759–1777.

Boukabara, S.A., T. Zhu, S. Lord, S. Goodman, R. Atlas, M. Goldberg, T. Auligne, B. Pierce, L. Cucurull, M. Zupanski, M. Zhang, I. Moradi, J. Otkin, D. Santek, B. Hoover, Z. Pu, X. Zhan, C. Hain, E. Kalnay, D. Hotta, S. Nolin, E. Bayler, A. Mehra, S.P.F. Casey, D. Lindsey, L. Grasso, V.K. Kumar, A. Powell, J. Xu, T. Greenwald, J. Zajic, J. Li, J. Li, B. Li, J. Liu, L. Fang, P. Wang, and T.-C. Chen, 2016: S4: An O2R/R2O infrastructure for optimizing satellite data utilization in NOAA numerical modeling systems: A step toward bridging the gap between research and operations. *Bulletin of the American Meteorological Society,* 97 (12), 2359–2378.

Hoffman, R.N., and R. Atlas, 2016: Future observing system simulation experiments. *Bulletin of the American Meteorological Society,* 97 (9), 1601–1616.

Hoffman, R.N., and R. Atlas, 2016: The OSSE checklist. *Bulletin of the American Meteorological Society,* 97 (9), ES193-ES199.

Kourafalou, V.H., Y.S. Androulidakis, G.R. Halliwell, H.-S. Kang, M. Mehari, M. Le Henaff, R. Atlas, and R. Lumpkin, 2016: North Atlantic Ocean OSSE system development: Nature Run evaluation and application to hurricane interaction with the Gulf Stream. *Progress in Oceanography,* 148, 1–25.

Lee, P., R. Atlas, G. Carmichael, Y. Tang, B. Pierce, A.P. Biazar, L. Pan, H. Kim, D. Tong, and W. Chen, 2016: Observing System Simulation Experiments (OSSEs) using a regional air quality application for evaluation. In *Air Pollution Modeling and Its Application XXIV,* D.G. Steyn and N. Chaumerliac (eds.). Springer International, 599–605.

Lee, S.-K., A.T. Wittenberg, D.B. Enfield, S.J. Weaver, C. Wang, and R. Atlas, 2016: U.S. regional tornado outbreaks and their links to the springtime ENSO phases and North Atlantic SST variability. *Environmental Research Letters,* 11 (4), 044008.

Ruf, C., R. Atlas, P. Chang, M.P. Clarizia, J. Garrison, S. Gleason, S. Katzberg, Z. Jelenak, J. Johnson, S. Majumdar, A. O'Brien, D. Posselt, A. Ridley, R. Rose, and V. Zavorotny, 2016: New ocean winds satellite mission to probe hurricanes and tropical convection. *Bulletin of the American Meteorological Society,* 97 (3), 385–395.

Halliwell, G.R., M. Mehari, L.K. Shay, V.H. Kourafalou, H. Kang, H.-S. Kim, J. Dong, and R. Atlas, 2017: OSSE quantitative assessment of rapid-response pre-storm ocean surveys to improve coupled tropical cyclone prediction. *Journal of Geophysical Research—Oceans,* 122 (7), 5729–5748.

Halliwell, G.R., M. Mehari, M. Le Henaff, V. Kourafalou, I. Androulidakis, H. Kang, and R. Atlas, 2017: North Atlantic Ocean OSSE system: Evaluation of operational ocean observing system components and supplemental seasonal observations for improving coupled tropical cyclone prediction. *Journal of Operational Oceanography,* 10 (2), 154–175.

Hoffman, R.N., S.-A. Boukabara, V.K. Kumar, K. Garrett, S.P.F. Casey, and R. Atlas, 2017: An empirical cumulative density function approach to defining summary NWP forecast assessment metrics. *Monthly Weather Review,* 145 (4), 1427–1435.

Li, J., Z. Li, T.J. Schmit, P. Wang, W. Bai, and R. Atlas, 2017: An efficient radiative transfer model for hyperspectral IR radiance simulation and applications under cloudy sky conditions. *Journal of Geophysical Research-Atmospheres,* 122 (14), 7600–7613.

McNoldy, B., B. Annane, S. Majumdar, J. Delgado, L. Bucci, and R. Atlas, 2017: Impact of assimilating CYGNSS data on tropical cyclone analyses and forecasts in a regional OSSE framework. *Marine Technology Society Journal*, 51 (7), 7–15.

Pu, Z., L. Zhang, S. Zhang, B. Gentry, D. Emmitt, B. Demoz, and R. Atlas, 2017: The impact of Doppler wind lidar measurements on high-impact weather forecasting: Regional OSSE and data assimilation studies. In *Data Assimilation for Atmospheric, Oceanic and Hydrological Applications, Volume 3*, S.K. Park and L. Xu (eds.). Springer International, 259–283.

Zhang, S., Z. Pu, D. Posselt, and R. Atlas, 2017: An impact of CYGNSS ocean surface wind speeds on numerical simulations of a hurricane in observing system simulation experiments. *Journal of Oceanic and Atmospheric Technology*, 34 (2), 375–383.

Annane, B., B. McNoldy, S.M. Leidner, R. Hoffman, R. Atlas, and S.J. Majumdar, 2018: A study of the HWRF analysis and forecast impact of realistically simulated CYGNSS observations assimilated at scalar wind speeds and as VAM wind vectors. *Monthly Weather Review*, 146 (7), 2221–2236.

Blackwell, W.J., S. Braun, R. Bennartz, C. Velden, M. DeMaria, R. Atlas, J. Dunion, F. Marks, and R. Rogers, 2018: An overview of the TROPICS NASA Earth Venture mission. *Quarterly Journal of the Royal Meteorological Society*, 141 (S1), 16–26.

Boukabara, S.-A., K. Ide, N. Shahroudi, Y. Zhou, T. Zhu, R. Li, L. Cucurull, R. Atlas, S.P.F. Casey, and R.N. Hoffman, 2018: Community global Observing System Simulation Experiment (OSSE) package (CGOP): Perfect observations simulation validation. *Journal of Atmospheric and Oceanic Technology*, 35 (1), 207–226.

Boukabara, S.-A., K. Ide, Y. Zhou, N. Shahroudi, R.N. Hoffman, K. Garrett, V. Krishna Kumar, T. Zhu, and R. Atlas, 2018: Community Global Observing System Simulation Experiment (OSSE) package (CGOP): Assessment and validation of the OSSE system using an OSSE/OSE intercomparison of summary assessment metrics. *Journal of Atmospheric and Oceanic Technology*, 35 (10), 2061–2078.

Bucci, L.R., C. O'Handley, G.D. Emmitt, J.A. Zhang, K. Ryan, and R. Atlas, 2018: Validation of an airborne Doppler wind lidar in tropical cyclones. *Sensors*, 18, (12), 4288, 1–15.

Christophersen, H., R. Atlas, A. Aksoy, and J. Dunion, 2018: Combined use of satellite observations and Global Hawk unmanned aircraft dropwindsondes for improved tropical cyclone analyses and forecasts. *Weather and Forecasting*, 33 (4), 1021–1031.

Cucurull, L., R. Atlas, R. Li, M.J. Mueller, and R.N. Hoffman, 2018: An observing system simulation experiment with a constellation of radio occultation satellites. *Monthly Weather Review*, 146 (12), 4247–4259.

Hoffman, R.N., V.K. Kumar, S.-A. Boukabara, K. Ide, F. Yang, and R. Atlas, 2018: Progress in forecast skill at three leading global operational NWP centers during 2015–2017 as seen in Summary Assessment Metrics (SAMs). *Weather and Forecasting*, 36 (6), 1661–1679.

Leidner, S.M., B. Annane, B. McNoldy, R. Hoffman, and R. Atlas, 2018: Variational analysis of simulated ocean surface winds from the Cyclone Global Navigation

Satellite System (CYGNSS) and evaluation using a regional OSSE. *Journal of Atmospheric and Oceanic Technology*, 35 (8), 1571–1584.

Li, Z., J. Li, P. Wang, A. Lim, J. Li, T.J. Schmit, R. Atlas, S.-A. Boukabara, and R.N. Hoffman, 2018: Value-added impact of geostationary hyperspectral infrared sounders on local severe storm forecasts via a quick regional OSSE. *Advances in Atmospheric Sciences*, 35 (10), 1217–1230.

Lopez, H., R. West, S. Dong, G. Goni, B. Kirtman, S.-K. Lee, and R. Atlas, 2018: Early emergence of anthropogenically-forced heat waves in the western US and Great Lakes. *Nature Climate Change*, 8 (5), 414–420.

Tratt, D.M., J.A. Hackwell, B.L. Valant-Spaight, R.L. Walterscheid, L.J. Gelinas, J.H. Hecht, C.M. Swenson, C.P. Lampen, M.J. Alexander, S.D. Miller, D.S. Nolan, J.L. Hall, R.M. Atlas, F.D. Marks, L. Hoffman, and P.T. Partain, 2018: GHOST: A satellite mission concept for persistent monitoring of stratospheric gravity waves induced by severe storms. *Bulletin of the American Meteorological Society*, 99 (9), 1813–1828.

Weatherhead, E.C., B.A. Wielicki, V. Ramaswamy, M. Abbott, T.P. Ackerman, R. Atlas, G. Brasseur, L. Bruhwiler, A.J. Busalacchi, J.H. Butler, C.T.M. Clack, R. Cooke, L. Cucurull, S.M. Davis, J.M. English, D.W. Fahey, S.S. Fine, J.K. Lazo, S. Liang, N.G. Loeb, E. Rignot, B. Soden, D. Stanitski, G. Stephens, B.D. Tapley, A.M. Thompson, K.E. Trenberth, and D. Wuebbles, 2018: Designing the climate observing system of the future. *Earth's Future*, 6 (1), 80–102.

Zhang, J.A., R. Atlas, G.D. Emmitt, L. Bucci, and K. Ryan, 2018: Airborne Doppler wind lidar observations of the tropical cyclone boundary layer. *Remote Sensing*, 10 (6), 825.

Alaka, G.J., X. Zhang, S.G. Gopalakrishnan, Z. Zhang, F.D. Marks, and R. Atlas, 2019: Track uncertainty in high-resolution HWRF ensemble forecasts of Hurricane Joaquin. *Weather and Forecasting*, 34 (6), 1889–1908.

Anthes, R., S. Ackerman, R. Atlas, L.W. Callahan, G.J. Dittberner, R. Edwing, P. Emch, M. Ford, W.B. Gail, M. Goldeberg, S. Goodman, C. Kummerow, M.W. Mair, T. Onsager, K. Schrab, T. von der Haar, and J. Yoe, 2019: Developing priority observational requirements from space using multi-attribute utility theory. *Bulletin of the American Meteorological Society*, 100 (9), 1753–1793.

Cui, Z., Z. Pu, V. Tallapragada, R. Atlas, and C.S. Ruf, 2019: Impact of CYGNSS ocean surface wind speeds on numerical simulations of hurricanes Harvey and Irma (2017). *Geophysical Research Letters*, 46 (5), 2984–2992.

Li, Z., J. Li, T.J. Schmit, P. Wang, A. Lim, J. Li, F. Nagle, W. Bai, J. Otkin, R. Atlas, R.N. Hoffman, S. Boukabara, T. Zhu, W. Blackwell, and T. Pagano, 2019: The alternative of CubeSat based advanced infrared and microwave sounders for high impact weather forecasting. *Atmospheric and Oceanic Science Letters*, 12 (2), 80–90.

Lopez, H., S.-K. Lee, S. Dong, G. Goni, B. Kirtman, R. Atlas, and A. Kumar, 2019: East Asian monsoon as a modulator of U.S. Great Plains heat waves. *Journal of Geophysical Research-Atmospheres*, 124 (12), 6342–6358.

Mears, C.A., J. Scott, F.J. Wentz, L. Ricciardulli, S.M. Leidner, R. Hoffman, and R. Atlas, 2019: A near-real-time version of the Cross-Calibrated Multiplatform (CCMP) ocean surface wind velocity dataset. *Journal of Geophysical Research—Oceans*, 124 (10), 6997–7010.

Ryan, K., L. Bucci, R. Atlas, J. Delgado, C. Landsea, and S. Murillo, 2019: Impact of Gulf-streamIV dropsondes on tropical cyclone prediction in a regional OSSE system. *Monthly Weather Review*, 147 (8), 2961–2977.

2020–21

Halliwell, G.R., G.J. Goni, M.F Mehari, V.H. Kourafalou, M. Baringer, and R. Atlas, 2020: OSSE assessment of underwater glider arrays to improve ocean model initialization for tropical cyclone prediction. *Journal of Atmospheric and Oceanic Technology*, 37 (3), 467–487.

Mueller, M.J., A. Kren, L. Cucurull, R. Hoffman, R. Atlas, and T. Peevey, 2020: Impact of refractivity profiles from a proposed GNSS-RO constellation on tropical cyclone forecasts in a global modeling system. *Monthly Weather Review*, 148 (7), 3037–3057.

Zheng, X., R. Atlas, R.J. Birk, F.H. Carr, M.J. Carrier, L. Cucurull, W.H. Hooke, E. Kalnay, R. Murtugudde, D.J. Posselt, J.L. Russell, D.P. Tyndall, R.A. Weller, and F. Zhang, 2020: Use of observing system simulation experiments in the U.S. *Bulletin of the American Meteorological Society*, 101 (8), E1427–E1438.

Shen, B.-W., R. Pielke, X. Zeng, J.-J. Baik, S. Faghih-Naini, J. Cui, and R. Atlas, 2021: Is weather chaotic? Coexistence of chaos and order within a generalized Lorenz model. *Bulletin of the American Meteorological Society*, 102 (1), E148–E158.

Bucci, L.R., S.J. Majumdar, R. Atlas, G.D. Emmitt, and S. Greco, 2021: Understanding the response of tropical cyclone structure to the assimilation of synthetic wind profiles. *Monthly Weather Review* 149(6), 2031–2047.

Christophersen, H.W., B.A. Dahl, J.P. Dunion, R.F. Rogers, F.D. Marks, R.M. Atlas, and W.B. Blackwell, 2021: Impact of TROPICS radiances on tropical cyclone prediction in an OSSE. *Monthly Weather Review*, 49, 2279–2298.

Technical Reports and Other Publications

1974–79

Atlas, R., 1974: Development of a “nested mesh” model. *GISS Res. Rev.*, 38–40.

Atlas, R., 1974: The development of an advective mixed-layer ocean model. *GISS Res. Rev.*, 37.

Spar, J., and R. Atlas, 1974: Model atmospheric response to sea surface temperature variations. *GISS Res. Rev.*, 40–44.

Spar, J., R. Atlas, and E. Kuo, 1974: A 30-day forecast experiment with the GISS model and updated sea surface temperature. *GISS Res. Rev.*, 36.

Atlas, R., 1975: Numerical experiments to investigate the role of horizontal temperature advection in a predictive mixed layer ocean model. *GISS Met. Res. Rev.*, 78–85.

Atlas, R., and J. Miller, 1975: Climatological forecasts of sea-surface temperature with a mixed layer ocean model. *GISS Met. Res. Rev.*, 75–77.

Atlas, R., 1976: Evaluating the impact of satellite soundings on local weather forecasts. *Proc., NASA Weather and Cli. Bas. Sci. Rev.*

Quirk, W., and R. Atlas, 1977: The effect of increased horizontal resolution on synoptic forecasts with the GISS model of the global atmosphere. *Proceedings, Third Conference on Numerical Weather Prediction*, 92–99.

Atlas, R., 1978: A 100 percent chance of weather. *Newsday*, February 1978.

Atlas, R., 1978: Quantitative precipitation forecast from the GLAS model. *Atmospheric and Oceanographic Research Review, NASA Tech. Mem. 80253*, 14–19.

Atlas, R., 1978: The development of a computerized procedure for the prediction of severe local storm potential. *Atmospheric and Oceanographic Research Review, NASA Tech. Mem. 80253*, 20–24.

Atlas, R., R. Rosenberg, and M. Eaton, 1978: Synoptic evaluation of the impact of sounding data insertions and increased horizontal resolution on GLAS model forecasts. *Atmospheric and Oceanographic Research Review, NASA Tech. Mem. 80253*, 7–13.

Ghil, M., M. Halem, and R. Atlas, 1978: Model-dependence studies of the effects of sounding temperature assimilation on weather forecasting. *Atmospheric and Oceanographic Research Review, NASA Tech. Mem. 80253*, 3–6.

Halem, M., M. Ghil, R. Atlas, J. Susskind, and W. Quirk, 1978: The GISS sounding temperature impact test. *NASA Tech. Mem. 78063.*

Atlas, R., 1979: A comparison of GLAS SAT and NMC high resolution NOSAT forecasts from 19 and 11 February 1976. *NASA Tech. Mem. 80591.*

Atlas, R., 1979: Case studies of major DST-6 sounding impacts with the GLAS model. *Proceedings, Fourth NASA Weather and Climate Program Science Review, CP-2076*, 147–152.

Atlas, R., 1979: Comparison of DST and NMC objective analyses. *GARP Proj. Rept.*, 107, 140–171.

Atlas, R., M. Halem, and M. Ghil, 1979: Subjective evaluation of the combined influence of satellite temperature sounding data and increased model resolution on numerical weather forecasting. *Proceedings, Fourth Conference on Numerical Weather Prediction*, Oct. 29–Nov. 1, 1979, Silver Spring, MD.

Atlas, R., M. Halem, and M. Ghil, 1979: The influence of satellite temperature sounding data and increased model resolution on forecasting. *The GARP Programme on Numerical Experimentation*, Rept. No. 19, p. 25.

1980–89

Atlas, R., 1980: Applications of the TIROS-N sounding and cloud motion wind enhancement for the FGGE "special effort." *Proceedings, Eighth Conference on Weather Forecasting and Analysis*, June 10–13, Denver, CO.

Atlas, R., 1980: Evaluation of NESS soundings used in the FGGE special effort. *Proc., VAS Demonstration Sounding Workshop, NASA CP-2157*, 31–40.

Atlas, R., 1980: Synoptic evaluation of GLAS and NMC high resolution forecasts from 19 and 11 February 1976. *Atmospheric and Oceanographic Research Review, NASA Tech. Mem. 80650*, 32–41.

Atlas, R., 1980: TIROS-N sounding and cloud motion wind enhancement for the FGGE "Special Report." *Proceedings, International Conference on Preliminary Results of the First GARP Global Experiment*, June 23–27, Bergen, Norway.

Atlas, R., and R. Rosenberg, 1980: Supplementary notes on the development and verification of the automated forecasting method (AFM). *Atmospheric and Oceanographic Research Review, NASA Tech. Mem. 80650*, 42–46.

Atlas, R., M. Halem, and M. Ghil, 1980: The combined influence of satellite temperature sounding data and increased horizontal resolution on GLAS model forecasts. *Atmospheric and Oceanographic Research Review, NASA Tech. Mem. 80650*, 18–25.

Atlas, R., R. Rosenberg, and G. Cole, 1980: A comparison between an empirical technique for the prediction of cyclogenesis and the LFM II. *Proceedings, Eighth Conference on Weather Forecasting and Analysis*, June 10–13, Denver, CO.

Atlas, R., R. Rosenberg, and S. Palm, 1980: A case study of the growth of prognostic differences between GLAS model forecasts from SAT and NOSAT initial conditions. *Atmospheric and Oceanographic Research Review, NASA Tech. Mem. 80650*, 26–31.

Atlas, R., G. Cole, R. Rosenberg, S. Palm, and A. Pursch, 1980: Enhancement of TIROS-N sounding and cloud motion wind data for the FGGE special effort. *Atmospheric and Oceanographic Research Review, NASA Tech. Mem. 80650*, 14–17.

Kalnay-Rivas, E., W. Baker, M. Halem, R. Atlas, and D. Edelmann, 1980: GLAS experiments with FGGE II-b data. *Proceedings, International Conference on Preliminary Results of the First GARP Global Experiment*, June 23–27, Bergen, Norway.

Atlas, R., 1981: Preliminary results of an assessment of FGGE special effort data and its impact on GLAS model analyses and forecasts. *Proceedings, Fifth Conference on Numerical Weather Prediction*, November 2–6, Monterey, CA.

Atlas, R., 1981: The effect of varying amounts of satellite data, orographic and diabatic processes on the numerical prediction of an intense cyclone. *NASA Tech. Mem. 83907*, 7–10.

Atlas, R., 1981: The relative contributions of increased resolution in the data assimilation and in the forecast model to satellite data impact. *NASA Tech. Mem. 83907*, 11–13.

Atlas, R., 1981: The role of data, resolution, and physical processes in the numerical prediction of an intense winter storm. *Current Problems of Weather Prediction Symposium Volume*, June 23–26, Vienna, Austria.

Atlas, R., G. Cole, A. Pursch, and C. Long, 1981: Interactive processing of Seasat scatterometer data. *NASA Tech. Mem. 83907*, 49–51.

Atlas, R., A. Pursch, C. Long, G. Cole, and R. Rosenberg, 1981: Preliminary evaluation of the FGGE special effort for data enhancement. *NASA Tech. Mem. 83907*, 52–60.

Halem, M., E. Kalnay-Rivas, W. Baker, and R. Atlas, 1981: The state of the atmosphere as inferred from the FGGE satellite observing systems during SOP1. *Proceedings,*

International Conference on Early Results of FGGE and Large-Scale Aspects of Its Monsoon Experiments, January 12–17, Tallahassee, FL.

Shukla, J., R. Atlas, and W. Baker, 1981: Numerical prediction of the monsoon depression of 5–7 July 1979. *Proceedings, International Conference on Early Results of FGGE and Large-Scale Aspects of its Monsoon Experiments*, January 12–17, Tallahassee, FL.

Atlas, R., 1982: A numerical investigation of intense coastal cyclogenesis, *Research Activities in Atmospheric and Oceanic Modeling*, p. 5.44.

Atlas, R., 1982: Satellite data in global scale modeling. In *Satellite Soundings and Their Uses, Part III*, Workshop on Satellite Meteorology, July 19–23, 1982, Ft. Collins, CO.

Atlas, R., and R. Rosenberg, 1982: Numerical prediction of the mid-Atlantic states cyclone of 18–19 February 1979. *NASA Tech. Mem. 83992*, 53 pp.

Atlas, R., P. Woiceshyn, S. Peteherych, and M. Wurtele, 1982: Analysis of satellite scatterometer data and its impact on weather forecasting. *Oceans,* September 1982, 415–420.

O'Brien, J., R. Atlas, et al., 1982: Scientific opportunities using satellite wind stress measurements over the ocean. Nova University, *N.Y.I.T. Press*, 152 pp.

Atlas, R., and A. Pursch, 1983: Model sensitivity to low-level wind specification. *NASA Tech. Mem. 84983*, 22–24.

Atlas, R., and R. Rosenberg, 1983: Case studies of coastal cyclogenesis. *NASA Tech. Mem. 84983*, 13–17.

Atlas, R., W.E. Baker, M. Halem, and E. Kalnay, 1983: Seasat-A impact experiments. *NASA Tech. Mem. 84983*, 25.

Atlas, R., W.E. Baker, E. Kalnay, and M. Halem, 1983: Application of scatterometer data to global weather prediction. *Research Activities in Atmospheric and Oceanic Modeling.*

Atlas, R., W.E. Baker, M. Halem, E. Kalnay, and P.M. Woiceshyn, 1983: The impact of Seasat-A scatterometer data on GLAS model forecasts. *Proceedings, Sixth Conference on Numerical Weather Prediction*, June 6–9, Omaha, NE, 226–231.

Baker, W.E., R. Atlas, M. Halem, and J. Susskind, 1983: A case study of the sensitivity of forecast skill to data and data analysis techniques. *Proceedings, Sixth Conference on Numerical Weather Prediction*, June 6–9, Omaha, NE, 200–205.

Baker, W.E., R. Atlas, M. Halem, and J. Susskind, 1983: The effect of data and data analysis techniques on numerical weather forecasts. *NASA Tech. Mem. 84983*, 18–21.

Duffy, D., and R. Atlas, 1983: Significant wave height predictions utilizing Seasat scatterometer data., *EOS*, 64, 1062.

Kalnay, E., and R. Atlas, 1983: Global analyses of ocean surface fluxes using Seasat scatterometer winds. *EOS*, 64, 1069.

Kalnay, E., R. Atlas, W. Baker, and M. Halem, 1983: FGGE forecast impact studies in the Southern Hemisphere. *Proceedings, First International Conference on Southern Hemisphere Meteorology*, July 31–August 6, Brazil.

Kalnay, E., R. Atlas, W. Baker, D. Duffy, M. Halem, M. Helfand, and R. Hoffman, 1983: Scatterometer applications to numerical weather prediction. *NASA Tech. Mem. 85632*, III-37.

Atlas, R., 1984: The effect of physical parameterizations and initial data on the numerical prediction of the President's Day cyclone. *Proceedings, Tenth Conference on Weather Forecasting and Analysis,* June 25–29, Clearwater Beach, FL, 580–587.

Atlas, R., and J. Firestone, 1984: Evaluation of GLAS model precipitation forecasts for North America during SOP-1. *NASA Tech. Mem. 86219.*

Atlas, R., E. Kalnay, J. Susskind, W. Baker, and M. Halem, 1984: Simulation studies of proposed observing systems and their impact on numerical weather prediction. *NASA Tech. Mem. 86219.*

Atlas, R., E. Kalnay, J. Susskind, W.E. Baker, and M. Halem, 1984: Simulation studies of the impact of advanced observing systems on numerical weather prediction. *Proceedings, Conference on Satellite Meteorology/Remote Sensing Applications.* June 25–29, Clearwater Beach, FL, 283–287.

Atlas, R., W.E. Baker, E. Kalnay, M. Halem, P. Woiceshyn, and S. Peteherych, 1984: The impact of scatterometer wind data on global weather forecasting. *Frontiers of Remote Sensing of the Oceans and Troposphere from Air and Space Platforms,* 567–574.

Atlas, R., E. Kalnay, W.E. Baker, J. Susskind, D. Reuter, and M. Halem, 1984: Observing system simulation experiments at GLAS. *Data Assimilation and Observing System Experiments.*

Duffy, D.G., and R. Atlas, 1984: A regional numerical weather prediction experiment using Seasat-A scatterometer data. *Research Activities in Atmospheric and Oceanic Modeling,* 7, 3.3.

Duffy, D., and R. Atlas, 1984: The impact of Seasat scatterometer data on surface wind and wave predictions for the QE II storm. *Oceans,* 183–188.

Duffy, D., and R. Atlas, 1984: The utilization of Seasat-A data in numerical weather prediction models. *Proceedings, Conference of Satellite Meteorology Remote Sensing Applications,* June 25–29, Clearwater Beach, FL, 261–265.

Kalnay, E., and R. Atlas, 1984: Global analyses of surface wind and ocean surface fluxes using Seasat scatterometer data. *Oceans.*

Kalnay, E., R. Atlas, W. Baker, and M. Halem, 1984: FGGE data impact studies in the Southern Hemisphere. *NASA Tech. Mem. 86219.*

Kalnay, E., R. Atlas, W.E. Baker, and M. Halem, 1984: Impact of drifting buoys on GLAS model forecasts in the Southern Hemisphere during SOP-2. *Research Activities in Atmospheric and Oceanic Modeling,* 7, 1.8.

Kalnay, E., R. Atlas, M. Halem, and W. Baker, 1984: Forecast skill of drifting buoys in the Southern Hemisphere. *Proceedings, Tenth Conference on Weather Forecasting and Analysis,* June 25–29, Clearwater Beach, FL, 329–332.

Kalnay, E., R. Atlas, W. Baker, D. Duffy, M. Halem, and H. M. Helfand, 1984: SASS wind forecast impact studies using the GLAS and NEPRF systems: Preliminary conclusions. *NASA Tech. Mem. 86219.*

Peteherych, S., M.G. Wurtele, P.M. Woiceshyn, D.H. Boggs, and R. Atlas, 1984: First global analysis of Seasat scatterometer winds and potential for meteorological research. *Frontiers of Remote Sensing of the Oceans and Troposphere from Air and Space Platforms*, 575–586.

Susskind, J., R. Atlas, and A. Pursch, 1984: High resolution GLAS retrievals on the McIDAS. *NASA Tech. Mem. 86219.*

Atlas, R., E. Kalnay, W.E. Baker, J. Susskind, D. Reuter, and M. Halem, 1985: Doppler lidar forecast impact study. *Third NASA/NOAA Infrared Lidar Back-scatter Workshop.*

Atlas, R., E. Kalnay, W.E. Baker, J. Susskind, D. Reuter, and M. Halem, 1985: Forecast impact simulation studies. *NASA Tech. Mem. 86219.*

Atlas, R., E. Kalnay, W.E. Baker, J. Susskind, D. Reuter, and M. Halem, 1985: Simulation studies of the impact of future observing systems on weather prediction. *Proceedings, 7th AMS Conference on Numerical Weather Prediction.*

Halem, M., 1985: Simulation studies related to the design of post-FGGE observing systems. *International FGGE Workshop*, Woods Hole, MA, July 2–20, 1984.

Kalnay, E., and R. Atlas, 1985: Analysis of ocean surface wind fields using Seasat-A scatterometer data. *NASA Tech. Mem. 86219.*

Kalnay, E., R. Atlas, W. Baker, and J. Susskind, 1985: GLAS experiments on the impact of FGGE satellite data on numerical weather prediction. *International FGGE Workshop*, Woods Hole, MA, July 2–20, 1984.

Wolfson, N., and R. Atlas, 1985: Objective determination of heat-wave patterns. *NASA Tech. Mem. 86219.*

Wolfson, N., R. Atlas, and Y. Sud, 1985: A numerical investigation of the summer 1980 heat wave and drought. *Proceedings, 7th AMÍ Conference on Numerical Weather Prediction.*

Atlas, R., 1986: Objective dealiasing and assimilation of Seasat scatterometer winds using the GLA GCM. *Research Activities in Atmospheric and Oceanic Modeling.*

Atlas, R., 1986: Simulation studies of the effect of low level wind data on Southern Hemisphere analysis and numerical forecasts. *Research Activities in Atmospheric and Oceanic Modeling.*

Atlas, R., and J. Pfaendtner, 1986: The effect of initial data, horizontal resolution, and oceanic fluxes on the numerical prediction of the Queen Elizabeth II storm. *Research Activities in Atmospheric and Oceanic Modeling.*

Atlas, R., R. Rosenberg, J. Ardizzone, and J. Terry, 1986: Global objective dealiasing of Seasat scatterometer winds. *NASA Tech. Mem.*

Atlas, R., N. Wolfson, and Y. Sud, 1986: The influence of boundary forcing on the maintenance and breakdown of the summer 1980 U.S. Heat Wave. *Research Activities in Atmospheric and Oceanic Modeling.*

Atlas, R., A.J. Busalacchi, E. Kalnay, S. Bloom, and M. Ghil, 1986: Global surface wind and flux fields from model assimilation of Seasat data. *Proceedings, Second AMS Conference on Satellite Meteorology/Remote Sensing Applications.*

Atlas, R., 1988: Recent SASS data import studies at GLA. *Research Activities in Atmospheric and Oceanic Modeling.*

Atlas, R., and S.C. Bloom, 1988: Verification of satellite surface wind speed directional assignment using simulated data. *Research Activities in Atmospheric and Oceanic Modeling.*

Atlas, R., N. Wolfson, and Y.C. Sud, 1988: Numerical prediction experiments related to the summer 1980 U.S. heat wave. *Proceedings, Eighth Conference on Numerical Weather Prediction.*

Bloom, S., and R. Atlas, 1988: Assimilation of satellite surface wind speed data using the GLA analysis/forecast system. *Proceedings, Third AMS Conference on Satellite Meteorology/Remote Sensing Applications.*

Atlas, R., and S.C. Bloom, 1989: Assimilation of satellite surface wind speed data. *Research Activities in Atmospheric and Oceanic Modeling.*

Atlas, R., and S.C. Bloom, 1989: Global surface wind vectors resulting from the assimilation of satellite wind speed data in atmospheric general circulation models. *Oceans.*

Atlas, R., and S.C. Bloom, 1989: Model assimilation of space-based ocean surface wind speed data. *EOS.*

Atlas, R., and N. Wolfson, 1989: Sensitivity of heat wave predictions to initial data. *Research Activities in Atmospheric and Oceanic Modeling.*

Atlas, R., N. Wolfson, and J. Terry, 1989: Simulation of rawinsonde displacement in OSSE's. *Research Activities in Atmospheric and Oceanic Modeling.*

Bloom, S.C., and R. Atlas, 1989: Analysis of space-based ocean surface wind speed data at GLA. *Use of Satellite Data in Operational Numerical Weather Prediction,* ECMWF/EUMETSAT.

Jusem, J.C., and R. Atlas, 1989: A numerical investigation of the River Plate's cyclone of 29–30 May 1984. *Third Conference on Southern Hemisphere Meteorology and Oceanography.*

Jusem, J.C., and R. Atlas, 1989: Pressure tendency analysis of GLA model forecasts. *Research Activities in Atmospheric and Oceanic Modeling.*

1990–99

Atlas, R., 1990: Impact of different LAWS orbits on data assimilation. *Research Activities in Atmospheric and Oceanic Modeling.*

Atlas, R., and S.C. Bloom, 1990: Analysis of SSM/I wind speed data. *Research Activities in Atmospheric and Oceanic Modeling.*

Atlas, R., and S.C. Bloom, 1990: Assimilation of satellite surface wind speed data in the tropics. *International TOGA Scientific Conference.*

Atlas, R., et al., 1990: Estimates of air-sea fluxes and ocean surface fields. *U.S. WOCE Implementation Plan.*

Atlas, R., et al., 1990: The U.S. contribution to air-sea flux estimates. *U.S. WOCE Planning Report Number 15.*

Atlas, R., 1991: Simulation studies of the impact of satellite temperature and humidity retrievals. *Report of the First GEWEX Temperature/Humidity Retrieval Workshop.*

Atlas, R., and G.D. Emmitt, 1991: Implications of several orbit inclinations for the impact of LAWS on global climate studies. *Proceedings, AMS Second Symposium on Global Change Studies.*

Chou, S.-H., R.M. Atlas, and M.P. Ferguson, 1993: Evaporation estimates over the western tropical Pacific ocean from Special Sensor Microwave Imager. *Preprints, 20th Conference on Hurricane and Tropical Meteorology,* San Antonio, TX, Amer. Meteor. Soc., 611–614.

Chou, S.-H., R.M. Atlas, C.-L. Shie, and J. Ardizzone, 1994: A westerly wind burst and its impact on latent heat flux observed from Special Sensor Microwave Imager. *Preprints, 7th Conference on Satellite Meteorology and Oceanography,* Monterey, CA, Amer. Meteor. Soc., 596–599.

Chou, S.-H., C.-L. Shie, R.M. Atlas, and J. Ardizzone, 1995: Impact of the December 1992 westerly wind burst on evaporation determined from SSMI data. *Preprints, 21st Conference on Hurricane and Tropical Meteorology,* Miami, FL, Amer. Meteor. Soc., 535–537.

Chou, S.-H., C.-L. Shie, R.M. Atlas, and J. Ardizzone, 1995: The December 1992 westerly wind burst on evaporation determined from SSMI data. *Proceedings, TOGA 95 International Scientific Conference,* Melbourne, Australia, 2–7 April 1995.

Atlas, R., 1996: Application of SSM/I wind speed data to weather analysis and forecasting. *Preprint, 15th Conference on Weather Analysis and Forecasting,* 138–141, 1996.

Atlas, R., 1996: The impact of ERS-1 scatterometer data on GEOS and NCEP model forecasts. *Reprint, 11th Conference on Numerical Weather Prediction,* 99–101.

Atlas, R., 1996: The December 1992 westerly wind burst and its impact on evaporation determined from SSM/I data. *Reprint, Eighth Conference on Air-Sea Interaction and Symposium on GOALS,* J81–J84.

Atlas, R., 1996: Validation of NSCAT data at GLA. *Preprint, 15th Conference on Weather Analysis and Forecasting,* 135–137.

Chou, S.-H., C.-L. Shie, R.M. Atlas, and J. Ardizzone, 1996: Evaporation estimates over global oceans from SSM/I data. *Preprints, 8th Conference on Satellite Meteorology and Oceanography,* Atlanta, GA, Amer. Meteor. Soc., 443–447.

Chou, S.-H., C.-L. Shie, R.M. Atlas, and J. Ardizzone, 1996: The December 1992 westerly wind burst and its impact on evaporation determined from SSM/I data. *Preprints, 8th Conference on Air-Sea Interaction and Conference on the Global Ocean-Atmosphere-Land System (GOALS),* Atlanta, GA, Amer. Meteor. Soc., J81–J84.

Atlas, R., 1997: Preliminary evaluation of NASA scatterometer data and its application to ocean surface analysis and numerical weather prediction. *Reprint. Earth Observing Systems II, SPIE Proceedings Series,* Vol. 3117.

Chou, S.-H., C.-L. Shie, R.M. Atlas, and J. Ardizzone, 1997: Air-sea fluxes over global oceans derived from satellite data. Abstracts, *1997 Joint Assemblies of IAMAS and IAPSO,* Melbourne, Australia, JMP9-3.

Atlas, R., 1998: A tropical-like cyclone in the extratropics. *Report, International Centre for Theoretical Physics,* 1–59.

Atlas, R., 1998: Experiments to determine the requirements for lidar wind profile data from space. *Reprint, Earth Observing Systems III, SPIE Proceedings Series,* Vol. 3439, 79–89.

Chou, S.-H., C.-L. Shie, R.M. Atlas, and J. Ardizzone, 1998: Global air-sea turbulent fluxes retrieved from satellite data. Abstract, *1998 Conference of World Ocean Circulation Experiment—Ocean Circulation and Climate,* Halifax, Canada, 24–29 May, 1998, 59.

Atlas, R., 1999: Application of lidar winds to data assimilation and numerical weather prediction. *Proceedings, SPIE Symposium on Laser Radar Technology and Applications IV,* Orlando, FL, April 6–9, 1999. International Society for Optical Engineering, *SPIE,* Vol. 3707, 268–277.

2000–2009

Otterman, J., R. Atlas, J. Ardizzone, T. Brakke, S.-H. Chou, J.C. Jusem, M. Glantz, J. Rogers, Y. Sud, J. Susskind, D. Starr, and J. Terry, 2000: Extreme winter/early-spring temperature anomalies in central Europe. *Research Reports Yagellonian University,* 15, 207–219.

Atlas, R., 2005: Results of recent OSSEs to evaluate the potential impact of lidar winds. *Lidar Remote Sensing for Environmental Monitoring VI,* U.N. Singh (ed.). *Proceedings, SPIE,* Vol. 5887, 118–125.

Atlas, R., 2005: The impact of AIRS on weather prediction. *Algorithms and Technologies for Multispectral, Hyperspectral, and Ultraspectral Imagery XI,* S.S. Shen and P.E. Lewis (eds.). *Proceedings, SPIE,* Vol. 5806, 599–606.

Atlas, R., 2005: The impact of current and future polar-orbiting satellite data on numerical weather prediction at NASA/GSFC. *Applications with Weather Satellites,* W.P. Menzel and T. Iwasaki (eds.). *Proceedings, SPIE,* Vol. 5658, 132–143.

Susskind, J., and R. Atlas, 2005: Atmospheric soundings from AIRS/AMSU in partial cloud cover. *Algorithms and Technologies for Multispectral, Hyperspectral, and Ultraspectral Imagery XI,* S.S. Shen and P.E. Lewis (eds.). *Proceedings, SPIE,* Vol. 5806, 587–598.

Atlas, R., O. Reale. B.-W. Shen, and S.-J. Lin, 2006: The use of remotely sensed data and innovative modeling to improve hurricane prediction. *Algorithms and Technologies for Multispectral, Hyperspectral, and Ultraspectral Imagery XII,* S.S. Shen and P.E. Lewis (eds.). *Proceedings, SPIE,* Vol. 6233, 62330U, https://doi.org/10.1117/12.673221.

Atlas, R., O. Reale, J. Ardizzone, J. Terry, J.-C. Jusem, E. Brin, D. Bungato, and P. Woiceshyn, 2006: Geophysical validation of WINDSAT surface wind data and its impact on numerical weather prediction. *Atmospheric and Environmental Remote Sensing Data Processing and Utilization II: Perspective on Calibration/Validation Initiatives and Strategies,* A.H. Huang and H.J. Bloom (eds.). *Proceedings, SPIE,* Vol. 6301, 63010C, https://doi.org/10.1117/12.680923.

Atlas, R., O. Reale, J. Ardizzone, J. Terry, J.-C. Jusem, E. Brin, D. Bungato, and J.F. Le Marshall, 2007: Evaluation of WINDSAT surface wind data and its impact on ocean surface wind analyses and numerical weather prediction. *Preprints, 11th Symposium on Integrated Observing and Assimilation Systems for the Atmosphere, Oceans, and Land Surface (IOAS-AOLS)*, San Antonio, TX, January 14–18, 2007, Amer. Meteor. Soc., CD-ROM, 6 pp.

Atlas, R., and L.P. Riishojgaard, 2008: Application of OSSEs to observing system design. In *Remote Sensing System Engineering*, P.E. Ardanuy and J.J. Puschell (eds.). *Proceedings, SPIE*, Vol. 7087:708707, https://doi.org/10.1117/12.795344, 9 pp.

Atlas, R., J. Ardizzone, and R.N. Hoffman, 2008: Application of satellite surface wind data to ocean wind analysis. In *Remote Sensing System Engineering*, P.E. Ardanuy and J.J. Puschell (eds.). *Proceedings, SPIE*, Vol. 7087:708707, https://doi.org/10.1117/12.795371, 7 pp.

Atlas, R., R.N. Hoffman, J. Ardizzone, M. Leidner, and J.C. Jusem, 2008: A new cross-calibrated, multi-satellite ocean surface wind product. *Proceedings, International Geoscience and Remote Sensing Symposium*, Boston, MA, July 7–11, 2008. IEEE Geoscience and Remote Sensing Society, CD-ROM, 4 pp.

Leidner, S.M., J. Ardizzone, J. Terry, E. Brin, and R. Atlas, 2008: Impact of satellite-derived ocean winds on hurricane forecasting at global and regional scales. *12th Symposium on Integrated Observing and Assimilation Systems for the Atmosphere, Oceans, and Land Surface (IOAS-AOLS)*, New Orleans, LA, January 20–24, 2008. American Meteorological Society, Boston, 5 pp.

Miller, T.L., R.M. Atlas, P.G. Black, J.L. Case, S.S. Chen, R.E. Hood, J.W. Johnson, J. Jones, C.S. Ruf, and E.W. Uhlhorn, 2008: Simulation of the impact of new aircraft and satellite-based ocean surface wind measurements on H*Wind analyses. *12th Symposium on Integrated Observing and Assimilation Systems for the Atmosphere, Oceans, and Land Surface (IOAS-AOLS)*, New Orleans, LA, January 20–24, 2008. American Meteorological Society, Boston, 7 pp.

Miller, T.L., R. Atlas, P.G. Black, C.C. Hennon, S.S. Chen, R.E. Hood, J.W. Johnson, L. Jones, C.S. Ruf, and E.W. Uhlhorn, 2008: Simulation of the impact of new ocean surface wind measurements on H*Wind analyses. *Extended Abstracts, 28th Conference on Hurricanes and Tropical Meteorology*, Orlando, Florida, April 28–May 2, 2008. American Meteorological Society, Boston, 7 pp.

Tao, W.-K., D. Anderson, R. Atlas, J. Chern, P. Houser, A. Hou, S. Lang, W. Lau, C. Peters-Lidard, R. Kakar, S. Kumar, W. Lapenta, X. Li, T. Matsui, M. Rienecker, B.-W. Shen, J.J. Shi, J. Simpson, and X. Zeng, 2008: A Goddard multi-scale modeling system with unified physics. *GEWEX News*, 18 (1), 6–8.

Atlas, R., R.N. Hoffmann, and J. Ardizonne, 2009: A cross-calibrated multiple platform ocean surface wind data set. *Proceedings, SPIE Symposium on Ocean Remote Sensing: Methods and Applications*, San Diego, CA, August 2, 2009. International Society for Optics and Photonics, *SPIE*, Vol. 7459, 9 pp.

Atlas, R., R.N. Hoffman, J. Ardizzone, M. Leidner, and J.C. Jusem, 2009: Development of a new cross-calibrated, multi-platform (CCMP) ocean surface wind product. *Extended Abstract, 13th Conference on Integrated Observing and Assimilation*

Systems for the Atmosphere, Oceans, and Land Surface, Phoenix, AZ, January 11–15, 2009. American Meteorological Society, Boston, 5 pp.

Matsutani, M., R. Atlas, et al., 2009: Expanding collaboration in joint OSSEs. *Extended Abstract, 13th Conference on Integrated Observing and Assimilation Systems for the Atmosphere, Oceans, and Land Surface,* Phoenix, AZ, January 11–15, 2009. American Meteorological Society, Boston, 6 pp.

Prive, N., Y. Xie, T.W. Schlatter, M. Masutani, R. Atlas, Y. Song, J. Woollen, and S. Koch, 2009: Observing system simulation experiments for unmanned aircraft systems: Preliminary efforts. *Extended Abstract, 13th Conference on Integrated Observing and Assimilation Systems for the Atmosphere, Oceans, and Land Surface,* Phoenix, AZ, January 11–15, 2009. American Meteorological Society, Boston, 4 pp.

Shen, B.W., W.-K. Tao, J.-D. Chern, R. Atlas, and K. Palaniappan, 2009: Scalability improvements in the NASA Goddard multiscale modeling framework for tropical cyclone climate studies. *Proceedings, HPC (High Performance Computing) Asia & APAN (Asia-Pacific Advanced Network) 2009 International Conference and Exhibition,* March 2–5, 2009, Kaohsiung, Taiwan. National Center for High-Performance Computing, 249–256.

2010–16

Atlas, R., 2010: Application of remotely sensed wind measurements to ocean surface wind analyses. *Proceedings, 2010 International Geoscience and Remote Sensing Symposium (IGARSS),* Honolulu, HI, July 25–30, 2010. Institute of Electrical and Electronic Engineers, 3 pp.

Atlas, R., 2010: Impact of satellite surface-wind data on weather prediction. *SPIE News,* https://doi.org/10.1117/2.1201007.003120, 2 pp.

Atlas, R., 2010: Review of observing system simulation experiments to evaluate the potential impact of lidar winds on weather prediction. *Proceedings, 2010 International Geoscience and Remote Sensing Symposium (IGARSS),* Honolulu, HI, July 25–30, 2010. Institute of Electrical and Electronic Engineers, 4 pp.

Atlas, R., R.N. Hoffman, S.M. Leidner, and J. Ardizzone, 2010: Impact of satellite surface wind observations on ocean surface wind analyses and numerical weather prediction. *Proceedings, SPIE Atmospheric and Environmental Remote Sensing Data Processing and Utilization VI: Readiness for GEOSS IV,* August 1–5, 2010, San Diego, CA. International Society for Optics and Photonics, *SPIE,* Vol. 7811, 8 pp.

Leidner, S.M., J. Ardizzone, J.C. Jusem, E. Brin, R.N. Hoffman, and R. Atlas, 2010: Ocean-surface wind impacts on hurricane forecasting: Regional and global examples. *Proceedings, 14th Symposium on Integrated Observing and Assimilation Systems for the Atmosphere, Oceans, and Land Surface (IOAS-AOLS),* Atlanta, GA, January 17–21, 2010. American Meteorological Society, Boston, 5 pp.

Atlas, R., 2012: Observing system simulation experiments to assess the impact of remotely sensed data on hurricane prediction. *Proceedings, SPIE Symposium on*

Imaging Spectrometry XVII, San Diego, CA, August 12–16, 2012. International Society for Optics and Photonics, *SPIE,* Vol. 8515, 8 pp.
Atlas, R., 2013: Evaluating the impact of future remotely sensed data on hurricane prediction. *SPIE Newsroom,* http://spie.org/x102772.xml (https://doi.org/10.1117/2.1201308.005047), 2 pp.
Atlas, R.M., G.D. Emmitt, and T.S. Pagano, 2013: Observing system simulation experiments to evaluate the impact of remotely sensed data on hurricane track and intensity prediction. *Proceedings, SPIE Symposium on Imaging Spectrometry XVIII,* San Diego, CA, August 25–29, 2013. International Society for Optics and Photonics, *SPIE,* Vol. 8870, 8 pp.
Atlas, R.M., 2014: Observing system simulation experiments to assess the potential impact of proposed satellite instruments on hurricane prediction. *Proceedings, SPIE Symposium on Imaging Spectrometry XIX,* San Diego, CA, August 17–21, 2014. International Society for Optics and Photonics, *SPIE,* Vol. 9222, 9 pp.
Atlas, R., 2015: Observing system simulation experiments to evaluate the potential impact of new remote sensing instruments on hurricane track and intensity forecasting. *Proceedings, SPIE Symposium on Imaging Spectrometry XX,* San Diego, CA, August 9–13, 2015. International Society for Optics and Photonics, *SPIE,* Vol. 9611, 10 pp.
Atlas, R., G.D. Emmitt, L. Bucci, K. Ryan, and J.A. Zhang, 2016: Impact of Doppler wind lidar data on hurricane prediction. *Proceedings, 18th Coherent Laser Radar Conference,* Boulder, CO, June 27–July 1, 2016. Cooperative Institute for Research in Environmental Sciences, 4 pp.

Index